REPORT

T0146375

Options for and Costs of Retaining C-17 Aircraft Production-Only Tooling

John C. Graser, Edward G. Keating,
Guy Weichenberg, Michael Boito,
Soumen Saha, Robert G. DeFeo, Steven Strain

Prepared for the United States Air Force

Approved for public release; distribution unlimited

RAND PROJECT AIR FORCE

The research described in this report was sponsored by the United States Air Force under Contract FA7014-06-C-0001. Further information may be obtained from the Strategic Planning Division, Directorate of Plans, Hq USAF.

Library of Congress Cataloging-in-Publication Data

Options for and costs of retaining C-17 aircraft production-only tooling / John C. Graser ... [et al.].
    p. cm.
  Includes bibliographical references.
  ISBN 978-0-8330-5889-8 (pbk. : alk. paper)
  1. C-17 (Jet transport)—Design and construction. 2. C-17 (Jet transport)—Costs. 3. Machine-tools—United States. I. Graser, John C.

  UG1242.T7O63 2012
  358.4'483—dc23

                                                                        2011049885

The RAND Corporation is a nonprofit institution that helps improve policy and decisionmaking through research and analysis. RAND's publications do not necessarily reflect the opinions of its research clients and sponsors.

RAND® is a registered trademark.

Published 2012 by the RAND Corporation
1776 Main Street, P.O. Box 2138, Santa Monica, CA 90407-2138
1200 South Hayes Street, Arlington, VA 22202-5050
4570 Fifth Avenue, Suite 600, Pittsburgh, PA 15213-2665
RAND URL: http://www.rand.org/
To order RAND documents or to obtain additional information, contact
Distribution Services: Telephone: (310) 451-7002;
Fax: (310) 451-6915; Email: order@rand.org

Preface

This report emanated from a RAND Project AIR FORCE study, "Options and Costs for C-17 Aircraft Tooling Retention." The goal of this study was to outline and assess options available to the U.S. Air Force (USAF) for preservation or disposal of unique tooling used to manufacture the C-17A aircraft. We also estimated the nonrecurring and recurring cost implications of someday restarting C-17 production. If C-17 production were restarted, the nonrecurring costs would be lower if the tooling had been retained. But tool retention can be costly and may not be a worthwhile investment, especially if a future production restart is unlikely.

This project was sponsored by Maj Gen Randal Fullhart, SAF/AQQ. The project's points of contact were Ronald Case and James Krieger of the C-17 division of the Mobility Directorate and Lt Col Trauna James of SAF/AQQM.

This research was conducted between June 2010 and March 2011. Related RAND documents include

- *Retaining F-22A Tooling: Options and Costs*, by John C. Graser, Kevin Brancato, Guy Weichenberg, Soumen Saha, and Akilah Wallace, TR-831-AF, 2011.
- *Ending F-22A Production: Costs and Industrial Base Implications of Alternative Options*, by Obaid Younossi, Kevin Brancato, John C. Graser, Thomas Light, Rena Rudavsky, and Jerry M. Sollinger, MG-797-AF, 2010.
- *Starting Over: Technical, Schedule, and Cost Issues Involved with Restarting C-2 Production*, by Obaid Younossi, Kevin Brancato, Fred Timson, and Jerry Sollinger, MG-203-Navy, 2003, Not available to the general public.
- *Reconstituting a Production Capability: Past Experience, Restart Criteria, and Suggested Policies*, by John Birkler, Joseph Large, Giles Smith, and Fred Timson, MR-273-ACQ, 1993.

This report should interest those involved in the acquisition of weapon systems, those concerned with the cost of such systems, and those interested in production shutdown decisions.

RAND Project AIR FORCE

RAND Project AIR FORCE (PAF), a division of the RAND Corporation, is the U.S. Air Force's federally funded research and development center for studies and analyses. PAF provides the Air Force with independent analyses of policy alternatives affecting the development, employment, combat readiness, and support of current and future air, space, and cyber forces.

Research is conducted in four programs: Force Modernization and Employment; Manpower, Personnel, and Training; Resource Management; and Strategy and Doctrine.

Additional information about PAF is available on our website:
http://www.rand.org/paf

Contents

Preface iii
Figures vii
Tables ix
Summary xi
Acknowledgments xv
Abbreviations xvii

CHAPTER ONE
Introduction 1

CHAPTER TWO
Tooling Issues 5
Sustainment Tools 6
Estimating Tool Values 7
Estimating Tool Sizes 8

CHAPTER THREE
Production Restart Costs 13
Data Sources and Assumptions 13
Nonrecurring Costs for Restarting C-17A Production at a Different Location ... 14
Nonrecurring Costs of a New Production Facility 15
Estimates of Remaining Nonrecurring Costs of a C-17A Restart 16
Recurring Costs of Restarting C-17A Production at a Different Location ... 19
Costs of Starting Up Production of a C-17 Variant at a Different Location ... 21
Recurring Costs of C-17 Variants 23

CHAPTER FOUR
Tooling Retention Analysis 25
Sensitivity of Findings to Restart Year Assumptions 30
Exploring Tool Obsolescence 32
More Tools Required for Sustainment 34
C-17 Variants 34

CHAPTER FIVE
Conclusions 39

APPENDIXES
A. A Model of Tooling Retention Desirability 43
B. A Comparison of C-17A and F-22 Tooling Retention 47

Bibliography 49

Figures

2.1. Size Categorizations of Other Fabrication Tools... 9
2.2. Relationship Between F-22 Tool Weight and Acquisition Cost............................ 11
4.1. A Tool Retention Decision Tree 25
4.2. Number of Production-Only Tools to Retain as a Function of Perceived Probability of C-17A Restart 29
4.3. Value and Differential Cost of Production-Only Tools to Be Retained as a Function of Perceived Probability of C-17A Restart 31
4.4. Diminishing Returns in Production-Only Tool Retention................................... 31
4.5. Number of Production-Only Tools to Be Retained with Different C-17A Restart Year Assumptions 32
4.6. Number of Production-Only Tools to Be Retained With and Without Annual 5-Percent Obsolescence 33
4.7. Number of Production-Only Tools to Be Retained with Different Numbers of Production-Only Tools... 35
4.8. Number of Production-Only Tools to Be Retained with Different Variants.............. 36
A.1. A Tool Retention Decision Tree with Model Parameters 45

Tables

S.1. Estimates of Nonrecurring Costs xii
2.1. Ownership Status of C-17A Tools 5
2.2. Estimated C-17A Sustainment Tools, by Type of Tool 7
2.3. Typical Dimensions and Weights of C-17A Tools 9
2.4. Acquisition Cost-Inferred F-22 Tool Sizes 11
2.5. Estimated Number of C-17A Tools, by Size 11
2.6. Estimated Number of C-17A Production-Only Tools, by Size 12
3.1. Key Assumptions for Production Restart Costs 14
3.2. Estimates of Nonrecurring New Facility and Tooling Costs of a C-17A Restart 15
3.3. Estimates of Remaining Nonrecurring Costs of a C-17A Restart 19
3.4. Estimates of Total Nonrecurring Costs of a C-17A Restart 19
3.5. Estimates of Recurring and Total Costs of a C-17A Restart 21
3.6. Estimates of Nonrecurring Costs of a C-17B Variant 23
3.7. Estimates of Nonrecurring Costs of a C-17FE Variant 23
3.8. Estimates of Recurring and Total Costs of C-17 Variants for 150 Aircraft
     Production 24
4.1. C-17A Tools' Assumed Compression Factors, by Size 27
4.2. Estimated Breakeven C-17A Restart Probabilities, by Size 27
4.3. Inputs to Assembly Tools' Estimated Breakeven C-17A Restart Probabilities,
     by Size 29
4.4. Small Other Fabrication Tool Breakeven Restart Probability as a Function of
     Year of Assumed Restart 33
4.5. Common, Rework, and New Percentage Estimates for C-17B Variant 35
4.6. Common, Rework, and New Percentage Estimates for C-17FE Variant 36

Summary

The U.S. Air Force (USAF) asked the RAND Corporation to analyze the desirability of storing government-funded, production-only tooling when production of the C-17A cargo aircraft ends. To address this question, we focused on weapon system–specific production-only tooling, i.e., tooling not used in weapon system sustainment and useful only for producing C-17s, that is, not readily convertible for use on a different weapon system.

Immediate disposal of weapon system–specific production-only tooling is usually the less costly option, but retention of this tooling gives the government the option of restarting production in the future without having to procure all-new tooling.

The possible restart scenarios include someday resuming C-17A production, starting up production of a tactical variant Boeing has proposed (which it refers to as the C-17B), or starting up production of the so-called C-17FE (*FE* standing for fuel efficient).

Future production of C-17As, C-17Bs, or C-17FEs is highly speculative. There is considerable uncertainty as to what sort of restart the USAF might want in the future, when, and in what quantities. Or, of course, C-17 production may never be restarted.

Tooling Issues

Boeing provided us a tally listing 53,910 government-funded tools currently used in C-17A production. These tools were distributed across nine tool types with more than half (31,025) being what Boeing terms *other fabrication tools*.

We assumed retention of any tooling currently being used in production that would also be needed for C-17A sustainment. Given that, of the total number of tools, we recommend retaining 9,761 for C-17A sustainment and that this set include all master models, hard masters, and stretch blocks (three of Boeing's nine types of tools) and/or their associated data.

The remaining 44,149 production-only tools appear to have little value for sustainment but may help reduce the cost of a prospective restart or future production of a variant. Thus, a key question became how much the tools would be worth in the event of a restart, in terms of the cost differential between retaining them and making new ones.

We assumed that, in case of a production restart, a tool would be worth its original acquisition cost, escalated into fiscal year 2011 (FY 2011) dollars, after adjusting for the cost of making the tool ready for production following a period in storage.

If we knew each tool's physical attributes, estimating the cost to pack, ship, and store that tool would be straightforward. Unfortunately, although the information Boeing provided us sorted each tool into one of the nine types based on its usage in production, it did not include the physical attributes of individual tools. To overcome this lacuna, we used the cost of each tool to associate it with one of three size gradations (small, medium, and large) within its tool type. Each of the resulting 27 categories was then assigned a typical weight and dimension based on discussions with Boeing experts. We estimated the costs to package, transport, and store each tool from these weight and dimension estimates.

Production Restart Costs

To assess how production restart costs would differ with and without retained C-17A production-only tooling, we analyzed three different scenarios: restarted C-17A production, a startup of C-17B production, and a startup of C-17FE production.

According to our estimates, the nonrecurring new facility and tooling costs for a C-17A restart would be about $1.4 billion (in FY 2011 dollars) with tooling retention and about $1.9 billion without it. This suggests that tool retention reduces nonrecurring tooling costs by about $540 million. Other nonrecurring costs for a C-17A restart, most centrally nonrecurring airframe engineering labor, would cost somewhere between $760 million and 1.34 billion. In total, therefore, the nonrecurring costs for a C-17A restart would be $2.1 billion to 2.7 billion with tool retention and $2.7 billion to 3.3 billion without it. See Table S.1.

Also, a production break leads to loss of learning, which imposes recurring cost penalties. These penalties would range from $8 million to 45 million per aircraft, with the largest penalty for a small restart quantity.

The costs for starting up production of a C-17 variant would be even higher (Table S.1). We estimated that the nonrecurring costs for a C-17B variant would be $4.6 billion to 6.4 billion with tool retention and roughly $450 million more without it. We estimated the nonrecurring costs for a C-17FE variant would be $6.2 billion to 7.0 billion with tool retention and roughly $300 million more without it.

We estimated that recurring costs for the C-17B would be slightly higher than those for the C-17A. Those for the C-17FE might be slightly lower or somewhat higher than those for the C-17A.

Ultimately, tooling costs are not a major cost driver. Tooling retention could reduce program acquisition unit cost by about 1.5 percent for a C-17A restart and about 1 percent for a variant startup.

Table S.1
Estimates of Nonrecurring Costs

Scenario	Estimate With Tool Retention		Increment Without Tool Retention ($M)
	Low ($B)	High ($B)	
C-17A restart	2.1	2.7	+540
C-17B startup	4.6	6.4	+450
C-17FE startup	6.2	7.0	+300

NOTE: All dollars in FY 2011 terms.

Tooling Retention Analysis

Clearly, the decision on retention must occur before determining whether to restart production. Other things equal, the higher the perceived probability of production restart, the greater the desirability of retaining production-only tooling. So, we developed a methodology for assessing the desirability of retaining C-17A production-only tooling.

We defined the *breakeven probability* of a production restart for a tool as the probability at which the decisionmaker is indifferent between retaining the tooling and not retaining it. If the decisionmaker's perceived probability of a restart is greater than the breakeven probability, he or she should retain the tooling and conversely. Breakeven probabilities are lowest, i.e., retention is most desirable, for high-value, low-volume tools that are inexpensive to retain but valuable at a restart.

Removing the master models, hard masters, and stretch blocks—all tools in these types are needed for sustainment—we estimated 18 different breakeven probabilities, three sizes for each of six types of tools. The lowest breakeven probability, of around 2 percent, is for the category of large other fabrication tools. On the other extreme, it would cost more to retain small handling fixtures and dollies and small workstands and storage racks than they are worth; tools in these two categories should not be retained even if restart is certain.

We are not prescribing or suggesting the actual probability of a C-17 restart. That subjective probability is a decisionmaker's choice. Conditional on making that choice, we have cataloged which categories of production-only tools should be retained and which should not.

Not surprisingly, there are diminishing returns on investments in tool retention. The first few millions of dollars of investments retain a considerable number of high-value tools. As more tools are retained, additional investments are less productive on the margin.

Tooling retention is more desirable when production restart comes sooner, although the optimal tool retention decision is only moderately affected by the restart year assumption. If tools' values decline while in storage, tooling retention is less desirable.

We assumed that a C-17FE would have less tool commonality with the C-17A than would a C-17B variant. Therefore, more tools should be retained if a C-17B startup is expected rather than a C-17FE startup.

Conclusions

Barring unforeseen changes to the C-17A program, production will end in 2014 or 2015. Once C-17A production in Long Beach ceases, any resumption of production would incur sizable costs. Even Table S.1's most optimistic C-17A restart case would have at least $2.1 billion in nonrecurring costs. The magnitude of the cost of restarting C-17A production or starting up production of a variant gives pause with respect to tooling retention. One could interpret these sizable cost estimates to suggest the probability of a future production restart is quite small. Without some probability of eventual C-17 restart, there would be no value in retaining C-17 production-only tools.

Our estimate of the nonrecurring cost of retaining production-only tools, net of the cost of near-term disposal, ranges from zero (if no production-only tools are retained) to about $70 million if nearly all tools for a C-17A restart were kept. To put tooling costs in perspective, if the entire population of C-17A production-only tools ($860 million worth) had to be reprocured for a restart of 150 C-17A aircraft, the program acquisition unit cost saving attributable to the retained tools would be about $6 million per aircraft or between 2 and 3 percent of the unit cost.

Acknowledgments

The authors appreciate the hard work of our Air Force points of contact, Ronald Case and James Krieger of the C-17 division of the Mobility Directorate and Lt Col Trauna James of SAF/AQQ. We also learned a great deal from Col Mark H. Mol (C-17 Program Manager) and George Labeau, Grant Lawless, and Gary Pence of the C-17 division of the Mobility Directorate, as well as James Pearson and Paul Phelps of the Warner Robins Air Logistics Center. Ford (Bill) Rowland of Air Mobility Command provided considerable project orchestration assistance. William Green of Air Mobility Command suggested the follow-up analysis that became Table 4.3.

Numerous contractor employees assisted us, including

- The Boeing Company
    - Long Beach, California: Craig Anderson, Alan K. Baker, Arthur J. Balazs, Randy Black, Thomas W. Butler, William J. Carolan, Felipe (Phil) Compean, John Dorris III, Ron Gill, Lisa Green, Gary J. Harris, Randy R. Peterson, Melvin E. Rice, Bob Roberts, Stephen A. Von Biela, and David B. Walker
    - Macon, Georgia: Matthew Grubb
    - San Antonio, Texas: Alphonso Aguilar, Joe Reeves, David Serna, Terri Snook, Dale Spidall, and William Thompson
    - St. Louis, Missouri: Ross Jacobs
- Ducommun, Gardena, California: Kenneth Stout
- Pratt & Whitney, East Hartford, Connecticut: Timothy J. McGuire, Robert F. Nowak, and Juergen W. Pfaff
- Triumph, Dallas, Texas: Douglas Rutter, Vic Sheldon, Allan Tepera, and Kenneth Wheeler.

We also learned about C-5B restart experiences from a series of interactions with Lt Gen Charles L. Johnson (USAF, ret.), now of The Boeing Company.

We received helpful reviews of an earlier version of this paper from our colleagues Fred Timson and Henry Willis. Carl Rhodes coordinated the document review process.

Program director Laura Baldwin provided comments on an earlier version of this report. We also thank our RAND colleagues Lisa Bernard, Michael Kennedy, Thomas LaTourrette, Kimbria McCarty, Megan McKeever, Christopher Mouton, David Orletsky, Deborah Peetz, Anthony Rosello, Hosay Salam, Jerry Sollinger, and Ellen Tunkelrott. Jane Siegel helped prepare this report. Phyllis Gilmore edited this report.

Our RAND colleague Michael Neumann's wife, Susan Diane Neumann, passed away while this report was being finalized. The authors extend our condolences to the Neumann family.

Of course, remaining errors are solely the authors' responsibility.

Abbreviations

AMP	Avionics Modernization Program
DMSMS	diminishing manufacturing sources and materiel shortages
DoC	Department of Commerce
DoD	Department of Defense
FY	fiscal year
JSTARS	Joint Surveillance Target Attack Radar System
MDS	mission design series
OMB	Office of Management and Budget
PAF	Project AIR FORCE
PBP	performance-based payment
PPTP	postproduction transition plan
R&D	research and development
RERP	Reliability Enhancement and Reengining Program
RTT	right-to-title
T-1	theoretical first unit
USAF	United States Air Force
USD(AT&L)	Under Secretary of Defense for Acquisition, Technology, and Logistics

CHAPTER ONE

Introduction

Constructing a modern weapon system, be it an airplane, ship, or ground vehicle, requires thousands of tools. These tools are used to fabricate the thousands of parts that go into the weapon system, to hold these parts in place during assembly operations, to calibrate the weapon system, and to otherwise help manufacturing workers produce the final product. Tools can vary widely in size and value, from small hand tools to massive assembly jigs and automated drilling and fastening machines.

When production of a weapon system ceases, a decision must be made as to what to do with this tooling. Some of it might be useful in supporting the weapon system throughout its life cycle. Some equipment and machines can be transferred to production or sustainment of other weapon systems. This report focuses on weapon system–specific production-only tooling: tooling not used in weapon system sustainment, useful only for producing a specific weapon system, and not readily converted for use on a different weapon system.

Department of Defense (DoD) policy supports two general types of tooling, based on the original source of funds used to procure it. The first category is funded directly by the government as part of the development or production contracts for a weapon system. Generally, tooling that is useful only for production of a specific weapon system (*unique tooling*) is funded by the government. Other tooling and equipment required for production but not unique to one weapon system (such as a robotic drilling machine or an overhead crane) is normally funded and procured by the weapon system contractor, which is then reimbursed for its use through depreciation charges included in its annual billing (*wrap*) rates. We refer to this type of tooling and equipment as *capital*. At the end of production, the contractor either retains capital equipment for other uses or disposes of it without further government involvement.

Government-funded tooling may, however, be kept for possible future production, converted to use for sustainment (such as being transferred to a depot or retained for future spare parts production), or disposed of.

The C-17A program involves three categories of government-funded tooling. The government directly owns some of it, even though Boeing or its suppliers use it in production. We refer to this as *government-owned tooling*. Other tooling is *right-to-title* (RTT) tooling, for which the government must assert its right to take title at the end of production if it wishes to retain the tool. Finally, there are performance-based payment (PBP) tools, whose ownership will revert to Boeing at the end of production. Federal Acquisition Regulation Part 52.232-32 discusses PBPs, including tool title vesting to the contractor on completion of its obligations under the contract.

We assumed, however, that the government must decide what to do with all tooling in any of these three categories. We ignored the possibility that the contractor might not relinquish the PBP tooling, rights to which revert to the contractor, and did not consider possible government payments to the contractor for PBP tooling if the government desired to retain the tooling.

Immediate disposal is usually the less-costly option, but government retention of production-only tooling gives the government the option of restarting weapon system production at some point without having to procure all-new tooling.

History indicates that weapon system production has not generally restarted once stopped. But, as Birkler et al. (1993) notes, there have been examples, such as the U-2 reconnaissance aircraft, the C-5B cargo aircraft, and the AGM-65 missile of the DoD paying to resume production that has ceased. Retaining production-only tooling could prove to be a low-cost investment with the potential of a large cost avoidance if a production restart did occur someday.

Production shutdown studies that address tool disposition, such as York et al. (1996), focus on estimating the cost of retention or disposal, conditional on a disposition decision being made. This report's analysis is complementary but instead focuses on the decision whether to save specific production-only tools.

We focus, in particular, on what should be done with C-17A production-only tools. The C-17A Globemaster III is a large transport aircraft developed by McDonnell Douglas (now part of Boeing) in the 1980s and early 1990s. The C-17A can transport equipment, supplies, and personnel over long distances, from one theater of operations to another, and can land on austere airfields with short runways (U.S. Air Force [USAF], 2008). In addition to the USAF, the C-17A is operated by the United Kingdom, Australia, Canada, the North Atlantic Treaty Organization, and Qatar. The United Arab Emirates has six aircraft on order, and Australia recently ordered a fifth C-17A. India has signed a contract for ten C-17As. Other foreign sales were uncertain as of May 2011.

USAF had taken delivery on 207 C-17As as of February 3, 2011. The current program of record funds a total of 222 USAF C-17As. The fiscal year 2010 (FY 2010) buy of ten aircraft was the last scheduled USAF buy. USAF and DoD leaders have stated there is no need for additional C-17As (see, for instance, Desjardins, 2010, and McCord, 2010). It seems unlikely that Congress will fund additional aircraft in the current budget environment.

Although the program had some troubled early years, the aircraft has become an Air Force workhorse, used heavily to support operations in Iraq, Afghanistan, and elsewhere. In fact, the C-17A has accumulated over 2 million flying hours with the second million recorded in just the last five years. With a stipulated design life of 30,000 flying hours, this would indicate the fleet has used up about one-third of its projected life. Continued high usage would call for a replacement or a significant service-life extension program in the next 20 years. If the frequency and intensity of overseas operations decrease and if C-17A use returns to a peacetime rate, the requirement for a replacement would extend further into the future.

As noted, when current production ceases, government-funded tooling must be dealt with. "Draping in place," or mothballing, is likely not an option unless a restart is imminent. Most centrally, when current production ends, Boeing seems likely to move out of the Long Beach, California, facility in which final assembly currently occurs. U.S. Government Accountability Office (2008) notes that Boeing officials have concluded that the Long Beach facility would not be used for future business and should be sold.

USAF asked RAND to help it determine what to do with this tooling by analyzing options for tooling retention, the cost of keeping the tooling, and the tooling's potential future value in a production restart.

There are different possible restart scenarios (without which one would never retain production-only tooling). Restarting production of C-17As is one possibility. Perhaps a missile-carrying C-17A could partially satisfy a long-range strike requirement.[1] Or perhaps deficiencies in a current cargo aircraft or changes in requirements could necessitate production of additional C-17As.

Another option is starting up production of the C-17B,[2] a variant Boeing has proposed that adds centerline landing gear, a tire deflation/inflation system, higher-thrust engines, advanced flaps, and an advanced situational awareness and countermeasures system.[3] The C-17B could be produced, for instance, to fulfill a tactical airlift requirement. It would have considerable commonality with the C-17A in terms of aircraft structure. Hence, a relatively large amount of the existing C-17A tooling could be used in C-17B production.

Boeing has also proposed a fuel-efficient variant, the C-17FE. This aircraft would have a narrower fuselage, up-rated engines, a double-element flap system, winglets, a longer loading ramp, a shorter cargo door, and a modified horizontal tail. Trade-off studies are ongoing to examine specific changes for the C-17FE. The C-17FE would have less tool commonality with the C-17A than the C-17B would have.

The Aeronautical Systems Center released and received responses to a Capability Request for Information related to a C-130 replacement. Such a program would likely begin development during the next decade. The C-17FE is a candidate for filling that role, but there would still be a production gap of perhaps eight to ten years between current C-17A production and a replacement program for the C-130. As we show in Chapter Four, the FE would utilize far fewer of the existing C-17A production-only tools than either the C-17A or B.

Future production of C-17As, C-17Bs, or C-17FEs is highly speculative. There is considerable uncertainty about what sort of restart the USAF might want in the future, when, and in what quantities. Or, of course, C-17 production may never be restarted.

The remainder of this report is structured as follows: Chapter Two discusses C-17A tooling issues. We note the different ownership structure of C-17A tools and note the important difference between sustainment and production-only tools. Chapter Three presents estimates of production restart costs, i.e., what the nonrecurring and recurring cost implications would be of someday restarting C-17A production or starting up production of a variant after a cessation of C-17A production. Retention of C-17A production-only tools would reduce the nonrecurring costs of restarting C-17 production, although only modestly in percentage terms, we found. In Chapter Four, we apply a model for tooling retention desirability to analyze which categories of C-17A production-only tooling are most desirable to retain. We estimate the minimum probability of a production restart required to justify retaining a category of production-only tooling. Not surprisingly, if an eventual production restart is thought to be more likely, it is recommended that more production-only tools be retained. Chapter Five presents conclusions.

---

[1]   AirLaunch, 2006, discusses experiments with releasing rockets from C-17As.

[2]   We use the vernacular *restart* to refer to resuming C-17A production in the future. We also use *restart* to refer generically to any resumption of C-17 production, be it of C-17As or a variant. We use the vernacular *startup* to refer to beginning C-17B or C-17FE production in the future.

[3]   The B and FE variants described in this report were Boeing-proposed design configurations as of December 2010, which we have used as a snapshot because such proposals evolve over time.

The report also has two appendixes. Appendix A provides the details of our model of tooling retention desirability. Appendix B compares the desirability of retaining C-17A production-only tooling to the desirability of retaining F-22 production-only tooling.

CHAPTER TWO

Tooling Issues

As noted in Chapter One, C-17A production uses different types of tooling and equipment. Table 2.1 enumerates the government-owned, RTT, and PBP tooling used in the production of the C-17A.

The government-funded tool tallies include tools used at Boeing sites as well as at their suppliers'. While there are many more RTT tools than tools in the other categories, the average cost per tool is similar between government-owned tools, RTT tools, and PBP tools.

There is other equipment used in the C-17A production process that is contractor-financed and therefore contractor-owned. Boeing refers to this equipment, most notably a number of Brotje automatic drilling and fastening machines, as *capital equipment.*

Additional tools, categorized as *nonaccountable* and *shop aids*, e.g., wire jig boards, are enablers in production that are owned by Boeing and therefore are not among the accountable tools. Some digital media tools, such as numeric control tapes, can be retained at virtually no cost. In this report, we do not analyze nonaccountable tools or digital media tools.

The varied ownership status of C-17A tools is worth noting in that, even if the USAF retained all the government-funded tooling, tooling gaps would remain that would need to be filled to restart production. Retaining tooling would reduce nonrecurring restart costs, but restart costs would remain sizable, as we discuss in Chapter Four.

Table 2.1
Ownership Status of C-17A Tools

Category	Quantity	Acquisition Costs (FY 2011)	
		Total ($M)	Average per Tool ($)
Government-funded tooling			
Government-owned	6,074	87	14,388
RTT	44,145	731	16,567
PBP	3,691	41	11,075
Total tools	53,910	860	15,946
Contractor-funded tooling			
Capital equipment		280	

NOTE: Capital equipment estimate is derived from a then-year dollar estimate provided by Boeing of its tools. Roughly $240 million of this total is at Long Beach. Data were not available for Boeing supplier capital equipment.

In our analysis, we considered prospective retention of all government-funded tooling. We did not estimate the potential consideration the government might be required to pay Boeing to obtain title to the PBP tools at their current locations. However, we considered the outside value of the PBP tools to be minimal, since all would generally be useless except for C-17 production. We also did not consider prospective retention of contractor-financed capital equipment but assumed that the contractors would make such equipment available for a C-17 production restart.

Boeing provided us tallies of the government-owned, RTT, and PBP C-17A tools falling into each of nine categories:

- master models: 413
- hard masters: 958
- very large complex tools: 66
- handling fixtures and dollies: 2,459
- workstands and storage racks: 504
- stretch blocks: 962
- other fabrication tools: 31,025
- assembly tools: 14,020
- special test equipment: 3,503.

The total is 53,910.

Sustainment Tools

We assumed that any tooling currently being used in production but also needed for C-17A sustainment would be retained. Given the information available to us, we recommend keeping 9,761 tools out of the 53,910 total for C-17A sustainment. Determining which tools will be needed to sustain the C-17A through its life cycle will require additional analysis, which should include an assessment of spares requirements and the tooling to produce them. Boeing and the USAF C-17 division of the Mobility Directorate, as part of the postproduction transition plan (PPTP) due to be completed in March 2012, will analyze sustainment requirements in depth. Table 2.2 gives the our best estimate of the quantities of C-17A sustainment tools based on information currently available, along with the percentage of tool type total.

We suggest retaining all master models and hard masters and/or their associated data because these are the sources for all parts and tools for the C-17A. Experts both within Boeing and in the USAF maintenance community concurred on the necessity to save the entirety of these two categories of tools.

We also suggest retaining stretch block tools required to make battle, crash, and accident damage repairs when metallic and composite skins are affected. The Warner Robins Air Logistics Center and the Boeing facility at the former Kelly Air Force Base (Boeing C-17 Sustainment Center, Port San Antonio) otherwise appear to have adequate tooling for other current maintenance and modification activities. Skin work is the exception to this generalization.

We urge retention of landing gear tools at Goodrich because of the number of observed cases of landing gear wear and tear and repairable crash damage on these aircraft. Without

Table 2.2
Estimated C-17A Sustainment Tools, by Type of Tool

Type of Tool	Quantity	Percentage of Tool Type Total
Master models	413	100
Hard masters	958	100
Very large complex tools	0	0
Handling fixtures and dollies	658	27
Workstands and storage racks	16	3
Stretch blocks	962	100
Other fabrication tools	5,177	17
Assembly tools	946	7
Special test equipment	631	8

these tools, there would be no manufacturing capability for replacement landing gear or parts at the maintenance depots.

We also urge retention of rotating tools for large assemblies. *Rotating tools*, often called *shipping fixtures*, are used to transport sections of an aircraft from one site to another. Without these, the aircraft section can be damaged in transit. While there are also disposable shipping fixtures made of plywood, rotating tools are usually sturdier and are often made of steel. Also, we urge retention of handling fixtures and hoist tools used in crash damage scenarios. Finally, we urge retention of all check fixtures for assemblies.

The remaining 44,149 *production-only* tools are the focus of this analysis. These tools appear to have little value for sustainment but may be able to reduce the cost of a prospective restart or future production of a variant. Thus, a key question is how much might they be worth if a restart occurs as compared to the differential cost of retaining them. *Differential cost* refers to the difference in cost between retaining a tool and disposing of it when current C-17A production ceases.

Estimating Tool Values

Ideally, to estimate the costs of keeping the production-only tools and their value in a production restart, we would know each tool's physical attributes, such as weight and dimensions, as well as its future value in a restart.

We received tool-specific acquisition costs from two Boeing government tool databases. We updated these costs to FY 2011 dollars using DoD price indexes. All dollar values in this analysis are in FY 2011 terms. However, not all Boeing suppliers were contractually obligated to report the acquisition dates of their tools. Boeing's data identified the tools without known acquisition dates as being acquired in 2001, the latest contract with which the tools were associated. The net effect is that some tools may be undervalued, i.e., they were actually acquired before 2001. But no data were available to us to address this problem or to calibrate its extent.

Given that the future value of a production-only tool is difficult to estimate, our baseline assumption is that a tool would be worth its original acquisition cost, escalated into FY 2011 dollars, in case of a production restart after adjusting for the cost of making it ready for production following a period in storage.

Perhaps this estimate of a tool's value is too high. Reasons to think a tool might be worth less than its FY 2011 acquisition cost in case of a production restart include

- A tool may have wear and tear when it goes into storage, implying it will not last as long or perform as well as a new tool upon emerging from storage. We assume poststorage refurbishment returns the tool to its prestorage status, but this cannot be expected to return a tool to like-new condition. Our conversations with Boeing production and tooling engineers indicated that most tools are not close to the end of their useful lives, however.
- The original acquisition cost for Boeing-manufactured tools included design costs that might not have to be borne if new, but identical, tools are used in a restart. The old design could be remanufactured from the stored digital design data, avoiding a redesign cost.
- Updated manufacturing techniques might call for different tools, so the retained tool may not be worth its original price due to technological obsolescence.

On the other hand, perhaps this estimate is too low. Reasons to think a tool might be worth more than its FY 2011 acquisition cost in case of a production restart include

- Finding a tool manufacturer at some future date that is still willing to make a legacy tool may result in a significant premium over the original acquisition cost.
- We believe some tools were acquired earlier than indicated in the Boeing data, in which case we have underescalated their acquisition costs into FY 2011 terms. For example, a tool that was acquired in FY 1991 should have a multiplier of 1.41 to escalate its acquisition cost into FY 2011 terms. However, if the data erroneously indicate it was acquired in FY 2001, we would only be using a multiplier of 1.21, thus understating the tool's value by 17 percent.
- The recorded tool acquisition cost does not include contractor and government acquisition costs that would be associated with buying a replacement tool.

Acknowledging each of these concerns, we know of no better proxy for a tool's value than its recorded acquisition cost escalated into FY 2011 dollars, less refurbishment or rework costs.

Estimating Tool Sizes

If we knew each tool's physical attributes, estimating the cost to pack, ship, and store that tool would be straightforward, e.g., the number of containers required.[1] Unfortunately, although the Boeing tooling databases sorted each tool into one of nine types based on its use in production, they did not supply the attributes of individual tools. To overcome this lacuna, we used the cost of each tool to associate it with one of three size gradations (small, medium, and large) within its tool type. Each of the resulting 27 categories was then assigned a typical weight and dimension based on discussions with Boeing experts. We estimated the costs to package, trans-

---

[1]   These would specifically be *Conexes* ("containers express"), reusable containers originally designed for shipping troop support cargo, quasi-military cargo, household goods, and personal baggage. Conexes, however, are now generally used for unit storage rather than deployment. See "Military Containers," undated.

port, and store each tool from these weight and dimension estimates. Table 2.3 presents the typical physical dimensions and weights by category used in our calculations.

Clearly, there is semantic confusion in dividing Boeing's *very large complex tools* into small, medium, and large. Even a small very large complex tool is larger than the large category for most types of tools. So the small, medium, and large labels should be understood to apply within the context of their types of tools, not as an absolute statement of their sizes.

We then used cost thresholds to assign each tool to its category's small, medium, or large gradation. Using our exploration of the data and consultation with Boeing experts, we decided to categorize any tool worth less than $10,000 as being small, tools worth $10,000 or more but less than $100,000 as being medium, and over $100,000 as being large. However, we created separate thresholds for the very large complex tool category, with the thresholds being

Table 2.3
Typical Dimensions and Weights of C-17A Tools

Type of Tool	Small		Medium		Large	
	feet	lbs	feet	lbs	feet	lbs
Master models	4 × 4 × 2	250	10 × 6 × 6	2,000	30 × 12 × 8	35,000
Hard masters	3 × 3 × 2	75	4 × 5 × 3	400	20 × 5 × 4	3,000
Very large complex tools	20 × 25 × 25	50,000	50 × 25 × 25	100,000	100 × 25 × 35	250,000
Handling fixtures and dollies	2 × 2 × 1	25	25 × 12 × 10	10,000	75 × 100 × 20	100,000
Workstands and storage racks	8 × 4 × 8	600	15 × 10 × 12	5,000	100 × 50 × 35	150,000
Stretch blocks	8 × 3 × 3	500	12 × 6 × 4	7,000	35 × 10 × 6	25,000
Other fabrication tools	2 × 1 × 1	5	4 × 4 × 2	50	15 × 5 × 4	300
Assembly tools	4 × 2 × 2	25	6 × 3 × 3	200	100 × 25 × 25	15,000
Special test equipment	2 × 2 × 1	20	5 × 3 × 4	150	12 × 12 × 15	3,000

Figure 2.1
Size Categorizations of Other Fabrication Tools

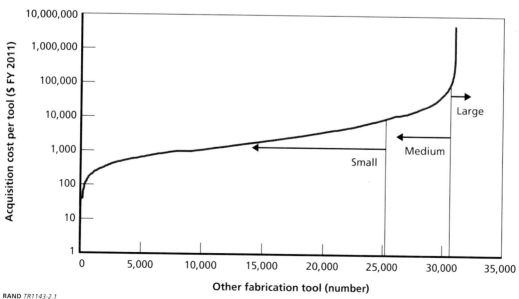

$2 million and $5 million. Boeing tooling experts concurred with the reasonableness of this approach.

For example, Figure 2.1 illustrates how we divided the 31,025 other fabrication tools into small, medium, and large. The 25,290 other fabrication tools with acquisition costs of less than $10,000 were categorized as small. The 5,349 other fabrication tools with costs between $10,000 and 100,000 were categorized as medium. The 386 other fabrication tools with acquisition costs more than $100,000 were categorized as large.

There is doubtlessly inaccuracy in these categorizations, e.g., some high-cost tools that might actually be small, some low-cost tools that are large. But without knowing individual tools' dimensions, we were forced into categorizing tools by using acquisition costs as a proxy for physical size. We felt categorizing all the tools into one of 27 categories was more accurate than using only a nine type-of-tool approach, which would assume, for instance, there is only one representative size of an other fabrication tool.

While we did not have individual C-17A tool sizes, we did have some individual Boeing tool weights in the F-22 tooling analysis, as described in Graser et al. (2011). As shown in Figure 2.2, there is a positive correlation between the weight of F-22 tools and their acquisition cost.

Figure 2.2 supports the notion of using tools' acquisition costs to infer their size. To assess the implications of our acquisition cost–size category procedure, we used the same $10,000 and 100,000 threshold technique to assign the F-22 tools to the three size gradations. Table 2.4 shows the results of those cost-based weight inferences.

It is heartening that average weight increased from the acquisition-cost-defined small-to-medium-to-large size gradations, consistent with the upward slope observed in Figure 2.2. Of course, a five-pound tool was not likely actually "large," but this was categorization error introduced by using acquisition cost to infer size. We have no way of knowing what the distributions of weights and sizes were in our C-17A tooling size graduations, although this F-22 exploration suggests the intrasize gradation dispersion is considerable. Thus, acquisition cost is only an imperfect proxy for a tool's weight or size. But we have no better approach for this C-17A analysis in lieu of actually measuring a large segment of the tooling population.

Our acquisition cost threshold technique distributed the 53,910 tools into 27 categories (see Table 2.5). Almost one-half of the tools fall in the small other fabrication tools category. Table 2.5 includes both production-only and sustainment tools.

We then subtracted the tools we identified previously as being required for sustainment from the total population to get the remaining 44,149 tools shown in Table 2.6. As discussed above, we believe all master models, hard masters, and stretch blocks are necessary to retain for sustainment purposes. Thus, none of the tools in these categories remains.

Figure 2.2
Relationship Between F-22 Tool Weight and Acquisition Cost

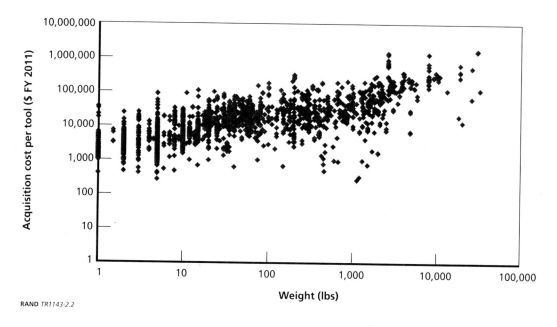

RAND *TR1143-2.2*

Table 2.4
Acquisition Cost-Inferred F-22 Tool Sizes

	Small	Medium	Large
Number of tools	870	1,109	226
Mean weight (lbs)	77	340	3,395
Standard deviation of weight	324	1,243	4,829
Minimum weight (lbs)	1	1	5
Maximum weight (lbs)	4,800	26,000	32,500

Table 2.5
Estimated Number of C-17A Tools, by Size

Type of Tool	Small	Medium	Large
Master models	133	217	63
Hard masters	486	374	98
Very large complex tools	27	33	6
Handling fixtures and dollies	1,654	744	61
Workstands and storage racks	165	234	105
Stretch blocks	449	424	89
Other fabrication tools	25,290	5,349	386
Assembly tools	11,281	2,339	400
Special test equipment	3,201	271	31

Table 2.6
Estimated Number of C-17A Production-Only Tools,
by Size

Type of Tool	Small	Medium	Large
Master models	0	0	0
Hard masters	0	0	0
Very large complex tools	27	33	6
Handling fixtures and dollies	1,222	542	37
Workstands and storage racks	159	227	102
Stretch blocks	0	0	0
Other fabrication tools	21,339	4,306	203
Assembly tools	10,616	2,077	381
Special test equipment	2,767	82	23

CHAPTER THREE

Production Restart Costs

We assessed how production restart costs would differ with and without retained C-17A production-only tooling. To do so, we analyzed three different scenarios: restarted C-17A production, a startup of C-17B production, and a startup of C-17FE production. We did not analyze a scenario Boeing has proposed of transitioning from C-17A production to C-17FE production, while keeping the Long Beach facility open throughout. Instead, the scenarios we focused on assume cessation of production in Long Beach followed by a restart some years later in a different location, e.g., Oklahoma. We compared restart or startup costs with and without retained tooling as part of the nonrecurring production cost estimate for each scenario.

An estimate of what a production restart might cost contributes to an assessment of the likelihood of such a restart someday occurring, an important parameter in Chapter Four's analysis.

Data Sources and Assumptions

The C-17 division of the Mobility Directorate provided us with information about C-17A airframe build quantities and negotiated costs. To date, all final assembly has taken place at the Boeing Long Beach facility. The C-17A's Pratt & Whitney F117 engines are provided to Boeing as government-furnished equipment.

For insight into C-17A costs at a different final assembly location, we used information on nonrecurring startup costs for facilities and tooling from U.S. Department of Commerce (DoC, 2005), the DoD facilities pricing guide (DoD, 2010), and RAND estimates. Boeing personnel knowledgeable about the C-17 program provided insights into the effects of a production break on the C-17 supplier base and efforts to restart production. We built estimates of recurring costs from the C-17 division-provided build quantities and negotiated costs, informed by Birkler et al.'s (1993) research on production breaks.

Our insights on the C-17B and C-17FE were informed by an enumeration of changes to the C-17A needed to manufacture the variants as of December 2010. The C-17A development and production programs were the primary sources of analogies.

Our estimates for aircraft development and recurring costs embody the assumptions enumerated in Table 3.1. Two of these assumptions, in particular, drive costs. For aircraft development, the estimates consist largely of labor effort, and costs therefore depend heavily on labor rates. We made the conservative assumption that labor rates in development would remain the same as at present. The potential move of Boeing final assembly to lower-cost areas and potential outsourcing of some design effort could decrease overall labor costs. For aircraft develop-

Table 3.1
Key Assumptions for Production Restart Costs

Subject	Assumption	Comment
Labor rates	Labor rates would not change.	Moving to a lower-cost area has the potential to reduce the cost of labor.
Disruption, loss of learning from production break	Stopping and restarting would incur significant penalties.	Perhaps vendor base could be preserved more than we assume.
Shutdown	The process would be "smart," including documention of current processes.	
Other Boeing facilities	Production activities at other locations would be available for a C-17 restart.	Only Long Beach final assembly facility will be replaced.
New final assembly site	A preexisting runway would be available at the final assembly site.	New runways are very expensive to build from scratch.
Final assembly	The facility would require 165 million ft^2 of space.	This is smaller than current Long Beach facility.

ment and production, the estimates depend heavily on assumptions regarding disruption to the supplier base and learning lost due to the break in production and relocation of final assembly. We have assumed that significant penalties are associated with stopping and restarting production, as we explain later in this chapter. If the C-17 vendor base were kept largely intact or reconstituted without significant loss of learning, restarted production costs could be lower than we estimate.

Additional important uncertainties affect the estimates presented in this chapter. There is no specific location for restarted production. The time frame and length of the hiatus in production are unknown. Details of the configuration of the variant of greater current interest to Air Mobility Command, the C-17FE, reflect a baseline as of December 2010, but changes in the configuration are likely.

Nonrecurring Costs for Restarting C-17A Production at a Different Location

We estimated the nonrecurring costs for restarting production activities currently performed at Long Beach (mainly final assembly and engineering support) at a different location, e.g., Oklahoma. Our restart cost estimates do not include the costs of shutting down the Long Beach facility or the eventual costs of shutting down a new facility.

For this exploration, we assume there is a "smart shutdown" and that the Long Beach facility is closed. In a smart shutdown, steps are taken to document production processes to minimize confusion in the event a different workforce undertakes a restart in the future. For example, videos and photographs of fabrication and assembly activities should be made, not only to record how the system was produced but to serve as training aids for follow-on workers. Interviews with key shop and technical personnel should be part of such documentation. See Birkler et al. (1993).

The nonrecurring cost estimate has two major elements. First, we estimated the costs of establishing a new production facility, including the cost of tooling and setting up the assembly line. Second, we estimated the remaining nonrecurring costs of restarting aircraft production, which are primarily labor costs. Many of these labor costs are for engineering effort to qualify or requalify vendors and to integrate and test new components. We estimated the recurring costs of aircraft production in light of the loss of learning entailed in moving from Long Beach to the new facility (see the next subsection). We assumed that production activities at other locations (e.g., Boeing plants in St. Louis, Missouri, and Macon, Georgia) would be available for a C-17 restart and that retained tooling could be reinstalled and production resumed.

Nonrecurring Costs of a New Production Facility

We estimated the costs of the new production facility primarily by applying costs per square foot from the DoD facilities pricing guide to estimates of the area required for current production at Long Beach. We used the cost estimates for utilities and infrastructure in the 2005 DoC study and RAND's estimate of current production-only tooling costs as the basis for the tooling estimate.

Table 3.2 presents estimates of the facility and tooling costs for a C-17A restart. Two RAND estimates are shown. The estimate for new tooling includes the cost of all-new production-only tooling.[1] The estimate for retained tooling assumes that all C-17A production tooling has been retained but that there are costs for refurbishing these tools and transporting them to the new location.

Our estimates in Table 3.2 and later tables may be measured in the tens of millions of dollars, but because of the considerable latent uncertainty in their calculation, they should not be seen as overly precise. Unfortunately, we were not able to assess a statistical range of uncertainty, such as a 95-percent confidence interval. Therefore, it is important to view our estimate of the total with new tooling in terms of approximation, for example, as *about* $1.9 billion" and that with retained tooling as *about* $1.4 billion." Tool retention, therefore, reduces nonrecurring tooling costs by *about* $500 million."

Table 3.2
Estimates of Nonrecurring New Facility and Tooling Costs of a C-17A Restart ($M, FY 2011)

Category	DoC Estimates	RAND Cost Estimates	
		New Tooling	Retained Tooling
Land	4		
Buildings	580	540	540
Utilities and infrastructure	100	100	100
Assembly line	390	280	280
Tooling	1,750	660[a]	120[b]
On-site design, engineering, and administration facilities	430	270	270
Off-site design, engineering, and administrative facilities plus equipment	450	80	80
Total	3,710	1,930	1,390

SOURCE: DoC, 2005, Table 5-2.

[a] This estimate presumes acquisition of all-new production-only tooling.

[b] This estimate presumes retention of all C-17A production tooling and its refurbishment and relocation.

---

[1] We further assumed that the production tooling retained to support sustainment requirements would be available and would not need to be reprocured.

DoC's estimate totaled about $3.7 billion in FY 2011 dollars (DoC, 2005, Table 5-2) and assumed replacement of all the original facilities. The DoC and RAND estimates diverge most in two specific categories. First, in terms of tooling, the RAND estimate reflects the value of currently used production tools. Boeing's current tool count is less than half the population used in the DoC study, which included all C-17A tools, not just government-funded tools. Second, for off-site design, engineering, and administrative facilities plus equipment, the RAND estimate assumes that these facilities (e.g., Boeing's facilities in St. Louis and Macon plus those of other subcontractors) would remain open for other uses and would not have to be reconstituted for a C-17 restart. The RAND estimate includes the cost of additional space for current off-site staff dedicated to the C-17.[2]

A number of other assumptions underlie the RAND estimates in Table 3.2. First, these estimates assume the new site is at a location with a preexisting runway, i.e., there is no incremental cost for building a runway, a very substantial cost to build from scratch.

The retained tooling estimate assumes the storage, relocation, refurbishment, and reuse of all production-only tooling, as well as relocation and resumption of use of sustainment tooling. The relocation of sustainment tooling would cost an estimated $5 million, included in the tooling total of $660 million in the "RAND Estimate with New Tooling" column. Likewise, the costs of relocating sustainment tools, as well as relocating and refurbishing retained production-only tools, are included in the $120 million tooling estimate in the retained tooling column.

The estimate for buildings assumes a required area for assembly of 1.65 million ft^2, which is less area than currently available because Bay 4 in Building 54 at Long Beach is not currently used for production. The cost per square foot came from the DoD facilities pricing guide for an aircraft maintenance hangar at an average continental U.S. location. We then used the utilities and infrastructure estimate from the DoC study for our own estimates.

The assembly line estimate is based on nonrecurring labor hours for the C-5B restart to do tool planning, manufacturing, quality assurance, and facilities engineering (Air Force Cost Analysis Agency, 1995; Lockheed Corporation, undated).

On-site design and engineering facilities were sized by the number of white collar employees at Long Beach as of July 2010. We assumed a total of 2,600 employees at 380 ft^2 of office space per employee multiplied by the DoD facilities pricing guide's estimate of the cost of aircraft research, development, test, and evaluation facilities. As mentioned, these estimates do not include costs to shut down Long Beach or, eventually, the new facility. Also, they do not include cost of capital equipment, assuming instead that Boeing would fund these costs and recover them through its overhead rates.

Estimates of Remaining Nonrecurring Costs of a C-17A Restart

Certain other types of nonrecurring costs for restarting C-17A production would be required independent of restarting in a new or the original location.

A major element of restarting C-17A production is the nonrecurring cost of engineering activities, including release of drawings, finding and qualifying or requalifying vendors, and integrating and testing new components when the component used on the current C-17A is

---

[2] On restart, other Boeing locations and subcontractors would incur costs for such things as training new workers and expanding or reconfiguring production facilities. We assumed that these costs would be reflected in the vendor costs for initial recurring production lots.

no longer available. To the extent that the original vendors and their equipment are available for restarted production, this effort would be minimal for a restart of the same configuration as the original program. In thinking through this portion of the estimate, we benefited from discussion with Boeing personnel on the potential effects of the cessation of C-17 production on the program's supplier base. The following paragraphs summarize the issues.

Initially, our thinking regarding this element was guided by the research of Birkler et al., which found that the nonrecurring engineering hours on the C-5B program were 8 percent of the nonrecurring labor on the C-5A program and proposed a range of nonrecurring labor hours on an S-3 restart of 2 to 9 percent of the original nonrecurring labor (Birkler et al., 1993, p. 13). The C-5B nonrecurring effort included engineering labor for configuration changes to the aircraft and thus included labor beyond what was required for tasks such as releasing drawings just to restart C-5 production. The C-5B was successfully restarted in 1982. Since that time, however, at least three things have changed that we believe would likely increase the resources needed to restart an aircraft program.

The first change is that the aerospace industry has experienced a trend toward outsourcing (offloading fabrication and subassembly work from the prime contractor to suppliers). Most of the value of the C-5B was produced in house in the mid-1980s. Today, most of the value of the C-17A is outsourced. Prime contractors today rely heavily on vendors and have less direct control over the means to build the parts used in the aircraft. With a prospective production hiatus, prime contractors would have less control over whether the capability to produce an outsourced part is retained and would have to find and qualify new vendors to replace those that no longer build parts for the program.

The second change since the 1980s is increased use of technology in aircraft. Cargo aircraft are often thought of as being relatively simple military aircraft, but one indication of the technological sophistication of the C-17A is that it requires over 2 million lines of code. The aircraft's avionics and other subsystems and the associated software interact continuously with each other. When components are changed, they must be tested to ensure that they work properly alone and be tested systemically to ensure that the entire architecture still works correctly. And because technology changes so rapidly, it is likely that many components would need to be replaced or modernized during a production restart.

The third change since the 1980s is faster evolution of technology. The consequences of rapid technological change and the products that embody it (especially in electronics) are recognized throughout DoD and the Air Force. Among these is that some items simply cease to be produced. The Air Force has a policy addressing diminishing manufacturing sources and materiel shortages (DMSMS) (AFMC Instruction 23-103, 2000), and major programs including the C-17A have DMSMS strategies that may include lifetime buys of items that are no longer going to be produced. The DMSMS problem is challenging and accounts for substantial funding for programs in production and sustainment. The problem would become much more difficult with a program restarted after a lengthy hiatus because the prime contractor would have to find and qualify or requalify vendors to make all the parts required for the C-17A and do so within a reasonable schedule before production could restart.

These changes suggest that more engineering labor could be required to restart C-17A production after a hiatus than the C-5B restart required, despite significant savings in tooling costs from the reuse of retained tooling. Unfortunately, there are no recent restart experiences that can serve as better analogies. We reflect the uncertainty in this element by presenting two estimates for nonrecurring labor restart costs. We expect that the length of the hiatus and the effects of ongoing C-17A modifications and DMSMS activities would affect the amount of nonrecurring engineering required at restart, and these factors are uncertain.

The lower estimate is guided by the C-5B experience cited in Birkler et al. (1993, p. 13), which indicates that C-5B nonrecurring engineering hours were 8 percent of C-5A nonrecurring engineering. Applying this percentage to C-17A nonrecurring engineering hours and using the labor rates in FY 2011 dollars that we assumed for a restart program, the lower estimate is $640 million.

Rough analogies from current modernization programs to the C-130 and C-5 that modify in-service aircraft with more modern and reliable equipment provided comparable and higher estimates. The C-130 Avionics Modernization Program (AMP) research and development (R&D) prime contract cost is $1.7 billion (FY 2011) for three C-130 mission design series (MDS) according to its December 2010 selected acquisition report, or roughly $570 million per MDS. This figure is slightly lower than the C-5B restart experience would suggest.

The C-5 AMP and Reliability Enhancement and Reengining Program (RERP) are two phases of a modernization effort for the C-5 that provides new engines, pylons, and several other subsystems. The R&D prime contract cost for the C-5 AMP is $480 million and the R&D prime contract cost for the C-5 RERP is $1.3 billion (both converted to FY 2011), according to the programs' December 2006 and 2009 selected acquisition reports, respectively. These modernization programs are crude analogies, especially in that they involve modifying the aircraft with equipment that improves capability, whereas the intent of the C-17A restart program would be to replicate existing capability.

Table 3.3 presents estimates of the remaining nonrecurring costs for restarting C-17A production. These are mostly labor costs.

The lower estimate assumes nonrecurring airframe engineering labor to release drawings, qualify and requalify vendors, and test and integrate new components and a level of effort comparable to the C-5B. The amount is close to one-third of what was spent for R&D for the three MDSs modified in the C-130 AMP. The higher, $1.3 billion, estimate assumes the need to replace and test more components, therefore requiring more greater engineering effort, and is based on the sum of C-5 AMP and RERP contractor R&D funding less the approximate amount of that funding for engine and airframe structural modifications.

The estimate for training new workers is three weeks for white collar employees and five weeks for production employees and is based on the methodology used in Younossi et al. (2010).

The cost to restart engine production assumes use of the same F117 engine and reflects an estimate of the cost for Pratt & Whitney to restart production of this engine. This estimate is based on an analogy to the Pratt & Whitney engine production restart associated with the Joint Surveillance Target Attack Radar System (JSTARS) program. Electronic Systems Center (2010) provided these estimates to RAND.

Table 3.4 draws together the estimates in Tables 3.2 and 3.3, summarizing our estimates of the nonrecurring costs of a C-17A restart. We reiterate our caveat about not inferring excessive precision from these estimates beyond "about $2.7 billion" and "about $3.3 billion."

Table 3.3
Estimates of Remaining Nonrecurring Costs of a
C-17A Restart ($M FY 2011)

	Estimates	
Category	Low	High
Nonrecurring airframe engineering labor	640	1,220
Training workers	80	80
Restart engine production	40	40
Total	760	1,340

Table 3.4
Estimates of Total Nonrecurring Costs of a C-17A Restart

	Estimate	
Category	Low	High
Nonrecurring new facility and tooling cost	1,390	1,390
Remaining nonrecurring costs of a C-17A restart	760	1,340
Total with tool retention	2,150	2,730
Increment for new tooling	540	540
Total with new tooling	2,690	3,270

NOTE: All dollars in FY11 millions.

Recurring Costs of Restarting C-17A Production at a Different Location

In estimating the recurring cost of restarted production, the main issue is the effect of the production break on cost improvement. Birkler et al. (1993) addresses this issue for labor hours. Its analysis of historical experience shows that cost-improvement curves are flatter and theoretical first-unit (T-1) costs are lower for a restart than for the original program.[3]

A key metric is the restarted T-1 as a percentage of the original T-1. Across different cost categories, these percentages varied, ranging from an engineering labor hours average of 29 percent to average production (manufacturing) and quality labor hours of 52 percent. For this estimate, we used a weighted average for the labor categories that resulted in a restarted T-1 cost of 48 percent of the original C-17A T-1 cost.

Birkler et al. (1993) also found that restarted learning-curve slopes were flatter, at 88 percent for a restarted program as opposed to 80 percent on the original program, on average. However, there was considerable variation in the restarted T-1 cost as a percentage of original T-1 cost and in learning-curve slopes across programs. The aircraft programs with restarts were the B-1, C-5, U-2, Jetstar, OV-10, and UH-2. The B-1, C-5, and UH-2 had production gaps lasting several years. Because the C-17A will be shutting down with no current requirement for additional strategic airlift identified by the USAF, we assumed the most likely production gap

---

[3] The designation *T-1* denotes the theoretical cost of the first unit produced. In practice, production costs are usually captured by production lot rather than by individual aircraft, so the actual cost of a given unit is seldom known. Unit costs are then estimated and designated *T*.

would be longer rather than shorter. (Our tooling retention analysis in Chapter Four assumes an expected production gap of 12.5 years, conditional on restart occurring.)

We also considered restarted production experience on two aircraft programs subsequent to Birkler et al. (1993). The Navy's E-2C program stopped production for two years (FYs 1993 and 1994) when it moved production from a facility in New York State to a new location in Florida. Also, the Navy's F/A-18E/F program stopped production for nearly two years after the flight-test aircraft were manufactured to avoid concurrency between development and production while flight testing was ongoing. For both programs, examination of cost data revealed little cost penalty due to the breaks in production.

However, the E-2C and F/A-18E/F analogies were unsuitable for our purposes because, in both cases, the programs were stopped with a known intent to resume production after a fairly brief period. Knowledge that production would resume made it easier to keep the programs' supply chains intact.

The C-17A case would be less favorable with no plans for restart upon cessation. Even if Boeing undertakes a smart shutdown and retains tooling for the program, that is not enough by itself because a large percentage of the value of the aircraft is produced by suppliers. Boeing personnel told us that several key suppliers would likely lose the capability to make parts or equipment for the C-17A after production ceased. Production restart would entail a schedule penalty to allow time to find and qualify new vendors and very likely a substantial cost penalty as the new vendors learned to manufacture the parts or equipment. Problems with even a few vendors could affect the schedule and production process for the entire program. The F-35 and Boeing 787 are examples of how aircraft production programs that rely heavily on outsourcing to new vendors can experience cost growth and schedule delays. See, for instance, Capaccio (2010) for a discussion of F-35 challenges and Gates (2011) for a discussion of 787 difficulties.

Using this reasoning, we chose a restart penalty in line with the averages reported in Birkler et al. (1993), rather than the more sanguine E-2C and F/A-18E/F experiences. We assumed that the restarted production would have a T-1 cost roughly one-half that of the original C-17A production and a learning curve slope roughly 6 percent flatter.[4]

Table 3.5 compares cost estimates for recurring (unit recurring flyaway) and total costs (program acquisition unit and total acquisition) associated with a C-17A restart to those for continuous production of the same number of aircraft in Long Beach. The program acquisition unit and total acquisition cost estimates that include nonrecurring costs assume retention of C-17A production-only tooling. The estimates in Table 3.5 reflect favorable assumptions for procurement and development costs. Because the main purpose of the table is to illustrate the difference between continuous and restarted costs, only the more favorable set of estimates are shown here.[5]

The top of Table 3.5 compares the costs of buying 25 more C-17As off the existing Long Beach production line with those for buying the same 25 additional C-17As off a restarted production line.[6] Measured in terms of recurring unit flyaway costs, there is a 21-percent penalty associated with restarted production. Adding the nonrecurring costs shown in Tables 3.3

---

[4]   The C-17A cost information was considered business sensitive and was provided to RAND with the understanding that it would not be disclosed.

[5]   See the last table in this chapter for lower and higher estimates that reflect more and less favorable assumptions.

[6]   We assumed that all cases of continuous production would begin from a low production rate sufficient to keep the Long Beach facility open at a notional five aircraft per year after current USAF buys are complete.

Table 3.5
Estimates of Recurring and Total Costs of a C-17A Restart

Buy Quantity	Production	Unit Recurring Flyaway	Program Acquisition Unit	Total Acquisition
25	Continuous (5-10-10)	214	233	5,830
	Restarted (5-10-10)	259	368	9,210
	Restart penalty	45	135	3,380
	Percentage	21	58	58
50	Continuous (5-10-15...)	203	221	11,070
	Restarted (5-10-15...)	231	295	14,730
	Restart penalty	28	74	3,660
	Percentage	14	33	33
100	Continuous (5-10-15...)	194	211	21,090
	Restarted (5-10-15...)	208	248	24,790
	Restart penalty	14	37	3,700
	Percentage	7	18	18
150	Continuous (5-10-15-...)	188	205	30,790
	Restarted (5-10-15-...)	196	228	34,250
	Restart penalty	8	23	3,460
	Percentage	4	11	11

NOTE: All dollars in FY 2011 millions.

and 3.4 to the program acquisition unit and total acquisition costs for the stop-and-restart approach results in a 58-percent penalty.

The relative penalty decreases as more aircraft are purchased. Table 3.5 illustrates this by showing buys of 50, 100, and 150 additional aircraft. At the highest restart quantity, 150, the recurring penalty is only 4 percent, while the total penalty is 11 percent.

Costs of Starting Up Production of a C-17 Variant at a Different Location

We also estimated costs associated with building a variant (either a C-17B or a C-17FE) at a new location. We did not evaluate the case of keeping Long Beach open and a transition from C-17A production to variant production. Instead, we assumed there would be a smart shutdown and that the Long Beach facility would be closed.

At Boeing's suggestion, we assumed that the C-17B and C-17FE would use Rolls Royce or different Pratt & Whitney commercial-derivative engines modified for the C-17 variant, while a C-17A restart would use the same Pratt & Whitney F117 engines currently used.

In addition to the general description of the variants provided in Chapter One, the key inputs for the parametric cost estimating methods we used to generate the production estimates in this chapter were airframe weight, material, and manufacturing processes. These details were provided to RAND as proprietary information and cannot be disclosed. In broad terms, the B variant is slightly heavier than the A, with very similar materials and production processes. The FE variant is a few thousand pounds lighter than the A, but with significantly greater use of composite materials and different production processes for many structural sections. The changes in the FE imply higher nonrecurring engineering and tooling costs than the B variant as well as different recurring production costs.

The three major elements of variant cost estimates are costs to establish a new production facility, costs of developing a C-17 variant, and recurring costs of aircraft production. We used the same methodologies as in the previous section to estimate the nonrecurring costs for a new production facility.

We then generated two estimates of development costs for C-17 variants. The first, lower estimate used cost-estimating relationships for airframe structural modification from Birkler and Large (1981). The second, higher estimate used the SEER-H model. SEER-H is a software tool that estimates development, procurement, and operating and support costs for new product development projects. For estimating aircraft costs, the model is sensitive to weight and material composition by aircraft section and to various features and processes in the design and production environment. Galorath Incorporated (2011) provides more information on SEER-H. Both methodologies use the amount of changed weight or design as a key input. For the C-17B variant, there was no change in avionics or other subsystems. But because of the potential for a hiatus to lead to changes in technology and therefore affect the availability of vendors and current components for restarted C-17 production, we added $600 million to the output of each methodology for the airframe estimate to reflect the nonrecurring effort associated with replacing obsolete or unavailable components.

Table 3.6 provides nonrecurring cost estimates for the C-17B including a new engine and the cost of a new production facility. Note that the estimates in Table 3.6 are in billions of dollars, while those in the earlier tables in this chapter were in millions of FY 2011 dollars.

The lower estimate, derived from the Birkler and Large (1981) methodology, began with an assumption of all-new tooling, then was decreased by the amount saved from refurbishing and reusing retained C-17A tooling applicable to the C-17B. We then added the costs to train new workers. The costs to restart and recertify a commercial-derivative engine, plus the costs to obtain maintenance drawings, rights to technical data, logistics planning, and initial planning to stand up depot support were based on analogy to the JSTARS reengining program. See USAF (2010, pp. 767–769). These costs were added to the airframe cost estimate. We also included the cost of new production facilities required to build the variant at a new location, for a total development cost estimate of $4.6 billion.

We calibrated the estimate using the SEER-H model to C-17A development costs from Boeing and applied the same tooling decrement and engine and new production facility costs. The estimated $4.6 billion to 6.4 billion in nonrecurring costs would increase by about $450 million, or roughly 7 to 10 percent, without retention of appropriate C-17A production-only tooling.

We likewise estimated the nonrecurring costs for a C-17FE variant, as presented in Table 3.7. The same methodologies and cost elements were applied. The development cost for the C-17FE variant is higher than that for the C-17B because there are more changes in the aircraft structure. Because the C-17FE design includes changes to avionics and other subsystems, both estimating methodologies generate some costs for these components, so we added only $300 million for the nonrecurring effort associated with replacing obsolete or unavailable components. Retaining C-17A tools is less valuable for production of the FE variant than of the B variant. The estimated $6.2 billion to 7.0 billion in nonrecurring costs would increase by about $300 million, or roughly 4 to 5 percent, without retention of appropriate C-17A production-only tooling.

Table 3.6
Estimates of Nonrecurring Costs of a
C-17B Variant ($B FY 2011)

	Estimate	
	Low	High
Airframe with all new tooling	3.2	5.0
Cost avoidance from refurbishment and reuse of C-17A tooling	(0.45)	(0.45)
Training new workers	0.1	0.1
Airframe total	2.9	4.7
Engine restart, certification, etc.	0.4	0.4
New production facilities	1.3	1.3
Total	4.6	6.4

Table 3.7
Estimates of Nonrecurring Costs of a C-17FE Variant ($B FY 2011)

	Estimate	
	Low	High
Airframe with all new tooling	4.7	5.5
Cost avoidance from refurbishment and reuse of C-17A tooling	(0.3)	(0.3)
Training new workers	0.1	0.1
Airframe total	4.5	5.3
Engine restart, certification, etc.	0.4	0.4
New production facilities	1.3	1.3
Total	6.2	7.0

Recurring Costs of C-17 Variants

We also estimated the recurring costs for restarting C-17 variants using a methodology derived from Younossi et al. (2001). We considered the variants' weights, material compositions, and fabrication processes. The empty weight of the C-17B is less than 1 percent greater than that of the C-17A, and the empty weight of the C-17FE variant is 5 percent less than that of the C-17A. Our estimated recurring labor hours at a common starting point for all variants, after adjustment for material composition and fabrication processes, are of similar proportions to the empty weights. Most importantly, we estimated where the different parts of the aircraft would be on cost improvement or learning curves.

We generated low and high recurring estimates, with the difference being due to where "new" or changed weight was placed on the learning curve and to the effect of a production break on unchanged components. The lower estimate assumed that only the structural weight that represented a significant change in configuration, such as the trailing flaps on the variants and the composite parts in the FE fuselage, was on the original C-17A learning curve, with its higher T-1 cost. All other weight was on the learning curve at a lower T-1 cost. This assumed that most of the changes in structure, many of which would involve changes to more automated fabrication processes, would not result in the higher manufacturing costs experienced in early C-17A production. It also assumed that vendors can be found readily for the unchanged components and that manufacturing of the unchanged items can be restarted without the higher costs experienced on initial C-17A production.

The higher recurring estimates assumed that all structural weight changes fell on the (less favorable) original C-17A learning curve. In addition, 25 percent of the unchanged non-structural weight for all variants, representing such subsystems as electrical, hydraulic, and other components, was assumed to be adversely affected by the hiatus, with greater problems manufacturing the unchanged items at prices experienced in the latter part of C-17A production. This weight also fell on the less favorable learning curve.

Table 3.8 provides estimates of the recurring and total costs associated with acquisitions of 150 C-17As, Bs, and FEs. Note that these estimates are for restarted production, not continuous production at Long Beach.

The lower estimate of recurring unit flyaway costs for the C-17A in Table 3.8 ($196 million) corresponds to that for a buy quantity of 150 in a restarted program in Table 3.5. The lower estimate of C-17A program acquisition unit costs in Table 3.8 ($228 million) corresponds to that for a buy quantity of 150 in a restarted program in Table 3.5.

In the more favorable case, we estimated that the recurring costs of C-17B production would exceed those for the C-17A by about $2 million per aircraft, while the recurring costs of C-17FE production would be about $1 million per aircraft less than our estimates for restarted C-17A recurring costs, assuming 150 aircraft were produced.

In the less favorable case, the recurring costs for C-17B production would be about $6 million per aircraft more than those for the C-17A, and the recurring costs for the C-17FE would be $9 million more than those for the C-17A.

The rightmost column in the table offers insight on the relative value of tooling retention. In the C-17A case, tooling retention would reduce nonrecurring costs by about $540 million (about $3.6 million per restarted aircraft, or about 1.5 percent of program acquisition unit cost). C-17B and C-17FE nonrecurring tooling cost avoidance of about $450 million and $300 million, respectively, would reduce program acquisition unit cost by roughly 1 percent. Tooling costs are not ultimately a major aircraft cost driver.

All the estimates in Table 3.8 would be higher if a smaller restart or startup quantity were chosen.

Table 3.8
Estimates of Recurring and Total Costs of C-17 Variants for 150 Aircraft Production ($B FY 2011)

Variant	Unit Recurring Flyaway		Program Acquisition Unit (includes development)		Tool Retention Savings Per Aircraft
	Low	High	Low	High	
C-17A	196	201	228	236	3.6
C-17B	198	207	246	269	3.0
C-17FE	195	210	254	276	2.0

CHAPTER FOUR

Tooling Retention Analysis

Appendix A details the methodology we use in this chapter to assess the desirability of C-17A production-only tooling retention for both a prospective C-17A production restart and a startup of C-17 variant production.

Figure 4.1 depicts the decision process. First, the Air Force must decide whether to retain a given tool. If a tool is retained, storage costs are incurred. If not, disposal costs are incurred. Later, production of the aircraft is either restarted or not. If a tool has been stored, a production restart implies the accrual of refurbishment costs and, eventually, disposal costs. If a tool has not been stored, a production restart implies the acquisition and eventual disposal of a replacement. If production does not restart, a stored tool must eventually be disposed of.

A key attribute of Figure 4.1 is that the tool retention decision precedes resolution of uncertainty about production restart. Therefore, the desirability of tooling retention is influenced significantly by the probability of a production restart. For the C-17, there is considerable uncertainty associated with this probability. (Chapter Three estimated the sizable costs that would be associated with a production restart.) Therefore, rather than attempt to estimate this important and highly uncertain parameter, we introduced the concept of the *breakeven probability of restart* to assist the Air Force in making its C-17A production-only tooling disposition decisions, which we defined as the probability at which the decisionmaker is indifferent between retaining or not retaining the tool. If the decisionmaker's perceived probability of a restart is greater than the breakeven probability, he or she should retain the tool. If the perceived probability of a restart is less than the breakeven probability, the tool should not be retained. Lower breakeven restart probabilities imply that tooling retention is more desirable,

Figure 4.1
A Tool Retention Decision Tree

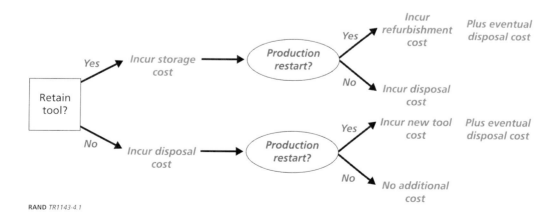

RAND *TR1143-4.1*

25

and vice versa. Appendix A contains a mathematical exposition of our breakeven probability of restart methodology.

Our breakeven probability approach is akin to what Posner (2005) terms "inverse cost-benefit analysis," i.e., dividing what the government is spending to prevent a catastrophic risk from materializing by what the social cost of the catastrophe would be if it did materialize. The result of this division is an approximation of the implied probability of the catastrophe. The Office of Management and Budget's (OMB's) Circular A-4 (OMB, 2003) explains that threshold or breakeven analysis answers the question, "How small could the value of the nonquantified benefits be (or how large would the value of the nonquantified costs need to be) before the rule would yield zero net benefits?"

A number of parameters are key to assessing a tool's breakeven probability. Some parameters affect how much it would cost to retain a tool:

- tool volume
- compression factor
- storage cost.

Others affect how much the tool would be worth if a restart occurred:

- tool cost
- refurbishment cost
- years in storage before restart.

To estimate retention costs, we assumed that tools would be shrink-wrapped at their current locations or at central collection points before being placed in crates (if required), then in 40 ft by 8 ft by 9.5 ft Conexes. Shrink wrapping tools ensures retention of the tool in near-original condition. We also assumed that some types of tools could be compressed before packing to reduce their dimensions. Table 4.1 presents our assumed compression percentages derived from expert judgments vetted with Boeing. According to these assumptions, items listed as 0 percent cannot be compressed, and items listed as 80 percent can be stored in a Conex taking up only 20 percent of its uncompressed volume. We arrived at these factors based on physical review of the tools and the presence (or absence) of logical joints and breakpoints for disassembly for storage.

The greatest compression for storage, we believe, can occur with handling fixtures and dollies and workstands and storage racks. For example, according to Table 2.3, the average large workstand and storage rack measures 100 ft by 50 ft by 35 ft, so its uncompressed cubic volume would be 175,000 ft². But since we assumed it could be compressed 90 percent, we further assumed that the Air Force could store it in a 17,500-ft² package.

We envisioned storage of tool-laden Conexes at the Sierra Army Depot in the open air. Leaving Conexes outdoors does not affect the condition of an enclosed tool significantly over time. If necessary, foam insulation can be installed in Conexes to ensure that moisture does not enter. We envision that radio frequency identification devices will be installed with all tools and Conexes.

Meanwhile, we used the acquisition costs of tools and estimated reuse costs (including refurbishment and retrieval costs) to calibrate the present value of the cost avoidance from retaining tooling in case of a restart. If retrieved, tool refurbishment includes cleaning, rust and oxidation removal, alignment, and testing. We assumed such refurbishment costs to be

15 percent of the acquisition cost of the tool, again based on expert judgment vetted with Boeing.

We assumed production restart is equally likely between years 5 and 20, with still-retained tools being disposed of after 20 years. We used data on the average dimensions of tools by category and retention and acquisition cost estimates to calculate category-by-category breakeven C-17A restart probabilities, shown in Table 4.2.

For example, we found that the production-only small tools in the very large complex tools category should be retained if there is at least an 18 percent probability of a C-17A restart.

In the table, the cells with checks indicate categories in which, as noted previously, we expect all tools to be retained for sustainment. We did not disaggregate retention decisions beyond the 18 unchecked categories in Table 4.2. Since we did not know individual tool sizes, we were forced to assume that a single decision would have to be made for each category, e.g., either all small other fabrication tools would be retained or all would be disposed of.

Table 4.1
C-17A Tools' Assumed Compression Factors, by Size (percent)

Type of Tool	Small	Medium	Large
Master models	0	0	0
Hard masters	0	0	0
Very large complex tools	50	60	80
Handling fixtures and dollies	80	85	90
Workstands and storage racks	80	85	90
Stretch blocks	0	0	0
Other fabrication tools	0	0	0
Assembly tools	0	40	70
Special test equipment	0	0	0

Table 4.2
Estimated Breakeven C-17A Restart Probabilities, by Size (percent)

Type of Tool	Small	Medium	Large
Master models	✓	✓	✓
Hard masters	✓	✓	✓
Very large complex tools	18	13	8
Handling fixtures and dollies	X	41	88
Workstands and storage racks	X	49	77
Stretch blocks	✓	✓	✓
Other fabrication tools	24	4	2
Assembly tools	67	8	63
Special test equipment	53	10	12

NOTES:
✓ Indicates tools we assume will be retained for sustainment regardless of any other decisions.
X indicates tools that our calculations suggest would cost more to retain than they are worth.

An X indicates categories in which our calculations suggest it would be more expensive to retain the tools than the tools are worth. The average small workstand and storage rack, for instance, is worth $3,804; however, given an average size 8 ft by 4 ft by 8 ft and weight of 600 lbs, it would likely have a nonrecurring storage cost of $7,849 by our calculations. Likewise, the average small handling fixture and dolly is worth $2,170 but a nonrecurring storage cost of $2,688. So, no restart probabilities make retaining the tools in these two categories worthwhile (even ignoring recurring costs of storage). While some individual small, high-value tools in these categories might be worth retaining, we could not undertake calculations at the individual tool level without knowing individual tools' sizes.

The tools in the two X'ed cells in Table 4.2 comprise 1,381 production-only tools worth about $3.3 million. We are not prescribing or suggesting the actual probability of a C-17 restart. That subjective probability is a decisionmaker's choice. Conditional on making that choice, Table 4.2 suggests which categories of production-only tools should be retained and which not.

Large other fabrication tools have the lowest breakeven restart probability in Table 4.2, about 2 percent. Based on a decisionmaker's probability of C-17A restart, he or she would choose to retain all the tools in the cells with breakeven restart probabilities less than the probability belief. So, for instance, a belief in a 10-percent probability of C-17A restart would imply retaining the large very large complex tools, the medium and large other fabrication tools, and the medium assembly tools and special test equipment. The rest should be disposed of after current production ceases.

Table 4.2 presented nonintuitive results for assembly tools. Both small and large assembly tools have fairly high breakeven restart probabilities (67 percent and 63 percent, respectively), while medium assembly tools have a much lower breakeven probability, 8 percent. Table 4.3 provides insight on the reasons for these results.

By construction, large assembly tools are more valuable, on average, than medium and small assembly tools. (This is not a finding; it is inherent to our cost-size categorization scheme.) Table 4.3 also repeats the tools' assumed typical dimensions and weight from Table 2.3, which came from Boeing. Table 4.1's compression factors were used to estimate average cubic foot to be stored in each category.

Medium assembly tools rate very highly on the average cost per cubic foot metric. While one takes up only roughly twice the space of a small assembly tool on average, it is more than ten times as valuable. Meanwhile, large assembly tools are, on average, more than ten times as valuable as medium assembly tools but take up nearly 600 times as much space. Results are similar, although not as dramatic, using average cost per pound.

Table 4.3 also presents the average nonrecurring storage cost for each size gradation. Nonrecurring storage costs are about 30 percent higher for medium assembly tools than for small assembly tools, while the medium tools cost more than 11 times as much on average. The ratio of nonrecurring storage costs to average cost is vastly lower, therefore, for medium assembly tools. Not surprisingly, therefore, medium assembly tools end up having far lower breakeven probabilities than small and large assembly tools.

It is important to note the methodological advancement latent in our approach. While Ebert (1992), for instance, outlined three prospective tooling disposition decisions (keep everything, dispose of everything, or keep only what might be needed for future support), we present here the more-nuanced option of retaining just the production-only tools that are most attractive to retain based on a comparison of their retention costs and value in case of a production restart.

Figure 4.2 plots the perceived probability of restart on the horizontal axis and the implied cumulative number of retained tools on the vertical axis. Not surprisingly, as the perceived probability of restart increases, the recommended number of production-only tools to retain also increases.

The figure is stepped because the analysis was done by category. If we had had more than 18 categories or, better yet, individual tool sizes, the function would have been smoother and less jagged. As the table suggests, if the perceived probability of restart of C-17A production were 10 percent, about 6,700 C-17A production-only tools should be retained. If the perceived

Table 4.3
Inputs to Assembly Tools' Estimated Breakeven C-17A Restart Probabilities, by Size

	Small	Medium	Large
Average cost ($ FY 2011)	2,410	28,114	286,816
Average dimensions (ft)	4 × 2 × 2	6 × 3 × 3	100 × 25 × 25
Compression factor (%)	0	40	70
Average cubic feet	16	32.4	18,750
Average weight (lbs)	25	200	15,000
Average cost per ft^3 ($)	151	868	15
Average cost per lb ($)	96	141	19
Nonrecurring storage cost	1,564	1,981	114,507
Nonrecurring storage cost/average cost	0.649	0.070	0.399
Estimated breakeven restart probability (%)	67	8	63

Figure 4.2
Number of Production-Only Tools to Retain as a Function of Perceived Probability of C-17A Restart

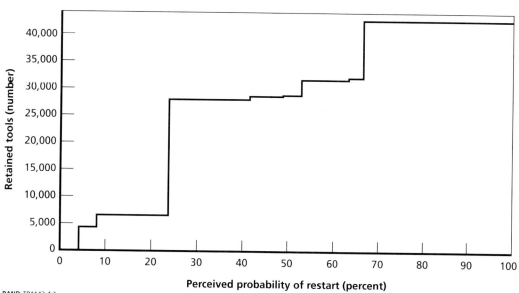

probability of restart were 25 percent, about 28,100 C-17A production-only tools should be retained.

The vertical axis in Figure 4.2 runs up to 44,149, reflecting the number of production-only tools in Table 2.6. However, even certainty of eventual production restart suggests retention of only 42,768 tools. The remaining 1,381 tools lie in the two X'ed cells in Table 4.2, suggesting that they are more expensive to retain than to replace.

Figure 4.3 plots the perceived probability of restart on the horizontal axis and the implied cumulative value of retained tooling on the left vertical axis, corresponding to the top line. Increasing the perceived probability of restart increases the value of tools retained. The middle line, corresponding to the right vertical axis, shows the differential total cost of retaining tooling as a function of the perceived probability of restart.

The lowest, broken line in Figure 4.3 shows the nonrecurring portion of the differential total costs of retention. This is an estimate of the funding that would be immediately required to retain the tools, ignoring the recurring costs of storage also built into the total cost of retention. For instance, if the perceived probability of restart of C-17A production were 10 percent, the figure indicates that $258 million worth of production-only tools should be retained. Doing so would have a total cost of about $10 million, with about half of that total in immediate nonrecurring costs. If the perceived probability of a C-17A restart were 25 percent, $445 million worth of tools should be retained at a total cost of $31 million, of which about $20 million would be an up-front, nonrecurring cost.

The left vertical axis in Figure 4.3 runs up to $655 million, reflecting our estimate of the total cost of production-only tools. However, even with certainty of eventual production restart, only $652 million worth of tools should be retained because the remainder fall in the two X'ed cells in Table 4.2, again suggesting that these tools are more expensive to retain than to replace.

Similarly, the right vertical axis runs to $139.5 million, the differential cost of retaining all C-17A production-only tooling. The table suggests that spending the $135.3 million to retain the $652 million worth of tooling would be optimal if restart were certain. The remaining $3.3 million worth of tooling would cost about $4.2 million to retain.

Figure 4.4 plots the differential cost of retaining tools on the horizontal axis and the associated value of tools that are optimally retained at this cost. Not surprisingly, there are diminishing returns in investments in tool retention. The first few million dollars of investments retain a considerable number of high-value tools. The more tools retained, however, the less productive the additional investments are on the margin.

Sensitivity of Findings to Restart Year Assumptions

In the remainder of this chapter, we undertake a series of robustness explorations examining how findings change under different assumptions and specifications, beginning with timing of the restart. In our base case, we assumed that production restart is equally likely anywhere between year 5 and year 20.

We experimented with two alternative parameterizations that preserved an expected restart year of 12.5. One case had restart equally likely in year 12 and year 13; the other case had restart equally likely in year 5 and year 20. These cases gave findings that were all but

**Figure 4.3
Value and Differential Cost of Production-Only Tools to Be Retained as a Function of
Perceived Probability of C-17A Restart**

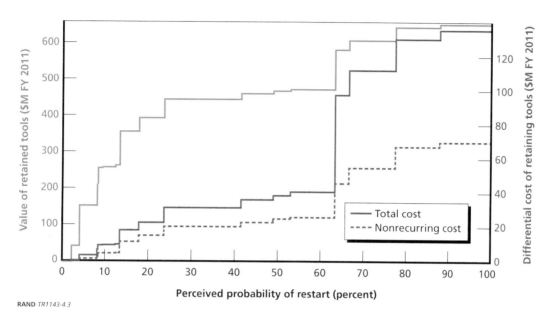

RAND TR1143-4.3

**Figure 4.4
Diminishing Returns in Production-Only Tool Retention**

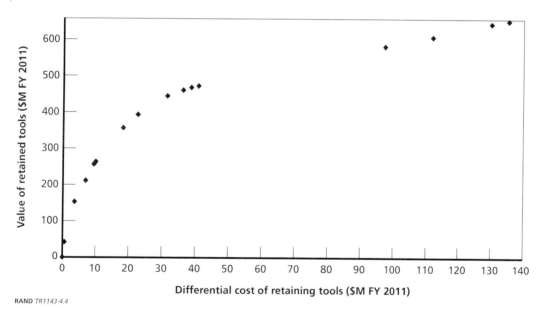

RAND TR1143-4.4

indistinguishable from our base case. The expected year of production restart drives total retention decisions without meaningful change related to the variance in the restart year.

Two alternative cases that gave moderately different findings were making the restart, if it
occurs, certain in year 5 and in year 20, respectively. As shown in Figure 4.5, restart in year 5
makes tooling retention more desirable; restart in year 20 makes tooling retention less desirable.

Restart in year 5 means fewer years of recurring storage costs, and the benefits of retention accrue sooner. The converse is true if restart occurs in year 20. Indeed, one surprise in Figure 4.5 is how moderately a major shift in the restart year assumption changes the optimal tool retention decision.

Table 4.4 illustrates the comparative lack of importance of the restart year assumption. This table presents different breakeven restart probabilities for small other fabrication tools. These breakeven restart probabilities vary only modestly whether restart is assumed to occur in year 5 or year 20. As expected, having restart sooner does reduce the breakeven restart probability.

Exploring Tool Obsolescence

Our base assumption was that a tool brought out of storage would be worth its initial acquisition cost less a reuse fee for bringing the tool to the new production facility and refurbishing it.

We explored an alternative parameterization in which we assumed a retained tool's value declined 5 percent annually in real terms. So, for instance, a tool worth $100,000 when put into storage would be worth about $61,000 (before payment of the reuse fee), in constant dollars, ten years later. Not surprisingly, such tool obsolescence tends to discourage tool retention, as shown in Figure 4.6.

The solid (no obsolescence) line in Figure 4.6 reprises the line in Figure 4.2 depicting the relationship between the perceived probability of restart and the optimal number of retained tools. The broken (annual 5 percent) line in Figure 4.6 plots the same probability—the number-of-tools relationship, but with the added feature of assumed 5-percent annual tool obsolescence (that a tool's value declines 5 percent annually in real terms). Fewer tools are retained in the presence of tool obsolescence.

Figure 4.5
Number of Production-Only Tools to Be Retained with Different C-17A Restart Year Assumptions

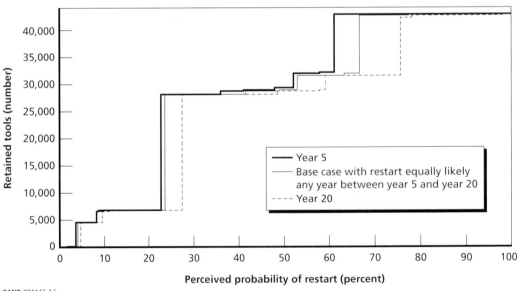

Table 4.4
Small Other Fabrication Tool
Breakeven Restart Probability as a
Function of Year of Assumed Restart

Assumed Restart Year	Breakeven Restart Probability (%)
5	22.7
6	23.0
7	23.3
8	23.5
9	23.8
10	24.1
11	24.4
12	24.7
13	25.0
14	25.3
15	25.7
16	26.0
17	26.3
18	26.7
19	27.0
20	27.4

Figure 4.6
Number of Production-Only Tools to Be Retained With and Without Annual 5-Percent Obsolescence

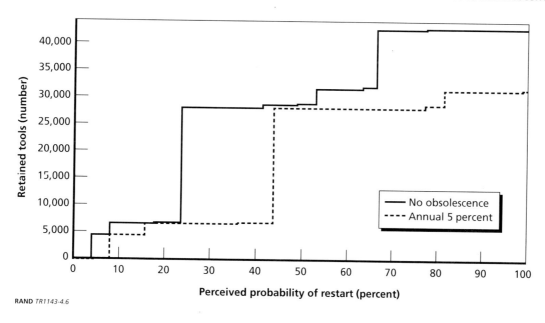

While we think there is an element of validity to growing tool obsolescence over time, we found no data that allowed calibration of any annual obsolescence rate. The 5-percent obsolescence rate we used in Figure 4.6 may be excessive and therefore understates the appropriate level of production-only tool retention. Of course, our no-obsolescence baseline assumption has the opposite bias, excessively favoring tool retention.

More Tools Required for Sustainment

We also explored the possibility that more tools would required for sustainment than we had previously estimated. In particular, Boeing experts thought our Table 2.2 sustainment tool tally of 9,761 tools might be too low. Boeing (2006) estimated that 39 percent of all production tools should be retained for sustainment. Unfortunately, a final sustainment tool count will not be available until completion of the ongoing PPTP, scheduled for March 2012.

To test the robustness of the analysis, we explored the possibility of the need for an additional 10,239 fabrication tools for sustainment, which would bring the new sustainment tool tally to 20,000 (equal to 37 percent of the total 53,910 production tools, thereby roughly equaling the Boeing, 2006, tool sustainment percentage). We assumed that all these additional sustainment tools would come from the other fabrication tools category because the extra tools would likely be required for fabrication of spares.

Figure 4.7 is an analog to Figure 4.2, except for the addition of the broken curve associated with the 33,910 production-only tools plus 20,000 sustainment tools as opposed to our baseline case of 44,149 production-only tools plus 9,761 sustainment tools. Not surprisingly, having more sustainment tools would make it necessary to retain fewer production-only tools. The overall shapes of the curves remain similar, with key decision points around the 25 and 65 percent probability levels.

C-17 Variants

We additionally assessed the usefulness of C-17A tools for the prospective B and FE variants. Ideally, a contractor would generate an engineering statement of work and a corresponding tooling statement of work. A *tooling statement of work* identifies specific tool numbers affected by the change, categorized into *common* (no change required), *rework* (poststorage rework would be needed for the tool to be used on the variant), and *new* (a tool would need to be entirely replaced).

However, absent such detailed statements of work, we assigned a percentage of common, rework, and new tools based on our discussions with and other information from Boeing concerning the C-17B and C-17FE variants. Our estimates are based on proposed B and FE design configurations as of December 2010.

We determined how many tools are used on each section of the aircraft using tool number nomenclature from Boeing. We also estimated what percentages of C-17A tools in each section were common (could be used with only refurbishment), required rework (minor modifications to allow use on the B or FE variants), or required new tools to be used on the variant. We assumed rework would cost 30 percent of original acquisition cost and rework would cost 15 percent, again based on expert judgment vetted with Boeing.

Table 4.5 presents our common, rework, and new estimated percentages for different sections of the C-17B aircraft.

Figure 4.7
Number of Production-Only Tools to Be Retained with Different Numbers of Production-Only Tools

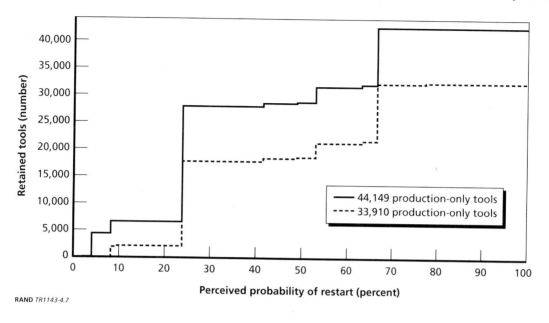

Table 4.5
Common, Rework, and New Percentage Estimates for C-17B Variant (percent)

Section of Aircraft	Common	Rework	New
Forward	80	10	10
Center	70	10	20
Aft	85	5	10
Vertical	90	5	5
Horizontal stabilizer	75	10	15
Wing	70	10	20
Pylon	90	5	5
Engine nacelle	80	10	10
General	80	10	10
Test	80	10	10

Table 4.6 provides parallel estimates for the C-17FE. The C-17FE has less tool commonality with the C-17A than does the C-17B. Most noticeably, since the C-17FE is presumed to use a new Rolls Royce or different Pratt & Whitney commercial-derivative engine, we assumed that all engine nacelle–related tools would need to be replaced.

Obviously, because it has more refurbishment-only and common tools and has fewer tools in need of replacement, tooling retention is relatively attractive for the C-17A.

Figure 4.8 reprises Figure 4.2's depiction of the optimal number of C-17A tools to retain relative to the perceived probability of restart. The two new lines correspond to C-17B and

Table 4.6
Common, Rework, and New Percentage Estimates for C-17FE Variant (percent)

Section of Aircraft	Common	Rework	New
Forward	75	15	10
Center	35	15	50
Aft	20	10	70
Vertical	25	15	60
Horizontal stabilizer	60	20	20
Wing	50	20	30
Pylon	30	30	40
Engine nacelle	0	0	100
General	50	25	25
Test	80	10	10

Figure 4.8
Number of Production-Only Tools to Be Retained with Different Variants

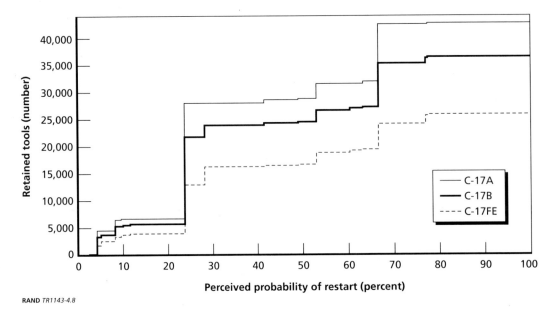

RAND TR1143-4.8

C-17FE startups, instead of a C-17A restart. It is most desirable to retain tooling if a C-17A restart is envisioned and least desirable if a C-17FE startup is envisioned (where fewer C-17A tools would be useful).

Figure 4.8 assumes that the decisionmaker decides *ex ante* that tools are being retained for a specific variant's restart or startup. So, for example, the figure's C-17FE line assumes that only tools that can be refurbished or reworked to be used in an FE startup are retained. It could turn out, of course, that the (*ex post*) "wrong" tools are retained. If, for instance, tools are retained envisioning an FE startup, but an A restart occurs instead, some tools the Air Force might wish it had retained will have been disposed of. Conversely, if tools are retained envisioning an A restart, but a B or FE startup occurs instead, some retained tools will be valueless.

A possible extension of this research would be to have three different subjective probabilities of restart, one for each prospective variant. This would involve separate analysis of the tools used solely on the A variant, tools used on the A and B variants, and tools used on all three variants. Or, although we do not believe it applies in this case, there could be tools used on the A and FE variants, but not on the B variant. Other things being equal, retention of tools valuable on any of the three variants would be most desirable. The most risk-averse approach would be to retain tools assuming a C-17A restart; this would be suboptimal if, for instance, C-17FE startup seemed disproportionately likely.

CHAPTER FIVE

Conclusions

Barring unforeseen changes to the C-17A program, production will end in 2014 or 2015. Senior officials in the OSD and the Air Force have repeatedly stated that the Air Force has enough intertheater airlift. With current pressures on the defense budget, add-on aircraft from Congress seem unlikely. India has committed to ten C-17s, and Australia has expressed interest in buying one more for its fleet. Other foreign sales are uncertain. Thus, if a decision is made to deviate from the current direction to the C-17 division of the Mobility Directorate to retain only sustainment tools, it will apparently have to be made in the next couple of years.

Retaining C-17A production-only tooling is valuable only if there is thought to be a chance of future resumption of C-17A production or starting up production of a variant.

Once C-17A production in Long Beach ceases, any resumption of production would incur sizable costs. In Chapter Three, we estimated that the nonrecurring cost of resuming C-17A production at a different location would be roughly $2.1 billion to 2.7 billion even with retention of all C-17A production-only tooling. The total nonrecurring cost would be over $500 million higher without tool retention. The nonrecurring costs for a C-17B startup would be even more considerable ($4.6 billion to 6.4 billion with tooling retention, and $450 million more without it), and the nonrecurring costs for a C-17FE startup might be even higher ($6.2 billion to 7.0 billion with tooling retention, and $300 million more without it). A restart or startup would also have sizable recurring learning curve cost implications.

The magnitude of the cost of restarting C-17A production or starting up variant production gives pause with respect to tooling retention. One could interpret these sizable cost estimates to suggest that the possibility of a future production restart is quite small.

Another major unknown in the tooling retention decision is obsolescence. Chapter Four showed the effect of gradual obsolescence of tooling using an illustrative annual rate of 5 percent. The net effect is that, under a certain C-17A production restart in 20 years, obsolescence would render approximately 30 percent of the tools not worth keeping. Notwithstanding that airframe technology does not evolve at a steady annual rate, the effects of obsolescence on the desirability of tool retention should be kept in mind because manufacturing processes and materials are all but certain to advance. Greater use of composites or other new materials, for example, will likely evolve over time, rendering some of the current C-17A tooling unusable or at least less desirable. In addition, despite continuous improvement of C-17A production processes over the past 20 years, the basic manufacturing approach is certainly not state of the art. Our discussions with C-17A production and engineering staff revealed that most felt that significant improvements could be made to the design and producibility of the aircraft in a restart if funding were allocated for such improvements. For example, determinate assembly could reduce the need for tools and reduce production costs.[1] Ironically, the large very large complex tool category, which we estimate has a relatively low breakeven probability (8 percent in Table 4.2), could be affected significantly by modern production approaches that emphasize the elimination of "monuments" in manufacturing processes.[2]

Another unresolved issue is how many C-17A production tools will be required to sustain current C-17A aircraft. Retention of production tools provides insurance against unforeseen sustainment requirements, such as crash or battle damage repairs. Several C-17As have received such damage. In some cases, tools were "borrowed" from the production line to complete repairs. We have identified approximately 9,800 tools which, coupled with 15 years of C-17A depot operations and the accumulation of 2 million flying hours, make totally unforeseen sustainment requirements unlikely. (The C-17A is a much more mature system than the F-22, whose tooling issues we discuss in Appendix B.) The ongoing Boeing and C-17 division of the Mobility Directorate PPTP will shed further light on sustainment tool requirements. If their analysis reveals the need for more tools than we have estimated, the lines depicted in the figures in Chapter Four will shift downward, as illustrated in Figure 4.7, as the production-only tool population shrinks. But while having more sustainment tools would reduce the costs of retaining production-only tools, it would axiomatically increase the cost of retaining sustainment tools. Our analysis has not estimated the nonrecurring costs associated with relocating sustainment tools from production facilities to depots or storage locations. The C-17 division of the Mobility Directorate's shutdown cost estimate will include such costs.

Our estimate of the nonrecurring cost of retaining production-only tools, net of the cost of near-term disposal of the tools, ranges from zero (if no production-only tools are retained) to about $70 million if nearly all tools for a C-17A restart are kept. (Recall that two categories of production-only tools would not be worth retaining even if eventual production restart were certain.)

To put tooling costs in perspective, if the entire population of C-17A production-only tools ($860 million worth) had to be reprocured for a restart of 150 C-17A aircraft, the program acquisition unit cost saving attributable to the retained tools would be about $6 million per aircraft or between 2 and 3 percent of the unit cost. Tooling is not a major cost driver in the total production cost of aircraft.

There are possible competitive implications to retaining C-17A production-only tools that our analysis did not consider. Retained C-17A tools are valuable only if C-17As or a variant are chosen to fulfill some future airlift requirement. In that sense, retention of these tools would provide an advantage to Boeing in some future competition. However, since tooling is not a major cost driver, that advantage would be slight. Nevertheless, one must acknowledge a potential scenario in which retention of C-17A production-only tooling discourages other firms from competing against Boeing to fulfill a future airlift requirement.

Finally, disposal of C-17A production-only tooling, which our findings tend to favor for at least some categories of tools regardless of estimated restart probabilities, would require a waiver from the Under Secretary of Defense for Acquisition, Technology, and Logistics

---

[1] Hartmann et al., 2004, describes determinate assembly as the practice of designing parts that fit together at a predefined interface and do not require setting gauges or other complex measurements and adjustments.

[2] *Monuments* is a pejorative term in lean manufacturing vernacular referring to "any design, scheduling, or production technology with scale requirements necessitating that designs, order and products be brought to a machine to wait in a queue for processing . . . monuments are waste." See Gembutsu Consulting LLC, 2009.

(USD[AT&L]). The FY 2009 National Defense Authorization Act, Public Law 110-417, Title VIII, Subtitle B, Section 815, requires that unique tooling associated with a major defense acquisition program be preserved and stored through the end of the service life of the weapon system. However, the Secretary of Defense can waive this requirement in the interest of national security, with notice to congressional defense committees.

An August 3, 2009, memo from USD(AT&L) (implemented effective March 2, 2011) notes that the office must be notified within 60 days if it is determined that preservation and storage of unique tooling is no longer in the best interest of the DoD. AT&L will then notify Congress.

Hence, this procedure will need to be followed if the decision is made to dispose of some or all of the C-17A production-only tooling.

A Model of Tooling Retention Desirability

This appendix provides a mathematical derivation of a tool's breakeven probability of restart. As mentioned in Chapter Four, the purpose of this concept is to assist the decisionmaker in making a tooling disposition decision on the basis of a comparison of his or her perceived probability of restart to an estimated threshold value. The breakeven probability of restart is the minimum perceived restart probability that renders tool retention more cost-effective than disposal.

Consider a production-only tool that, on cessation of production, can either be retained or disposed of. This disposition decision amounts to a comparison of

- the expected additional cost of retaining the tool
- the expected cost avoidance associated with retaining the tool instead of disposing of it.

A decision to dispose of the tool incurs an immediate nonrecurring cost, D_0. A decision to retain the tool incurs an immediate nonrecurring cost for packaging, transporting to a storage site, and putting the tool into storage, R_0. Tool retention also implies recurring storage costs. If the tool is stored for i years, the recurring storage costs total S_i. S_i sums multiple years' recurring costs, then discounts them into present value terms. Hence, for instance, if annual constant-dollar recurring costs in year t were x_t,

$$S_i = \sum_{t=1}^{i} \frac{x_t}{(1+d)^t},$$

where d is the real discount rate. OMB (2010) prescribes use of a 2.1 percent 20-year real interest rate for calendar year 2011, so we use $d = 0.021$ throughout this analysis. Therefore, the additional cost of retaining the tool for i years instead of disposing of it immediately is $R_0 + S_i - D_0$.

Typically, one will not know the value of i, the number of years of storage before restart occurs or storage is ended. We assert there is a maximum storage duration of N years after which tools still in storage will be destroyed. Experts we talked to, for instance, felt it reasonable to suppose tools still in storage after 20 years would no longer have any value, so we assume, in the analyses presented in Chapter Three, that $N = 20$. We further assume storage has some minimum possible duration we denote m. In our analyses in Chapter Four, we assume $m = 5$, i.e., the minimum amount of time tooling will spend in storage before a restart would be five years. The value of i is bracketed between m and N. Let w_i denote the probability restart occurs in year i, conditional on restart occurring. Then

$$\sum_{i=m}^{N} w_i = 1,$$

i.e., restart occurs somewhere between year m and year N.

Let p denote the overall probability of restart occurring. The probability restart occurs in any particular year i is $w_i\, p$. Should a restart occur in any of these years, there would be no differential cost of eventual disposal, as disposal of either the retained tools or the replacement tools will occur. With probability $1 - p$, no restart occurs, storage costs S_N are borne, and the retained tools are disposed of after N years at disposal cost D_N. In real dollars, disposal costs may be the same now as in N years. However, in present value terms, using the prescribed 20-year real discount rate of 2.1 percent, $D_N < D_0$. All our cash flows are in real FY 2011 terms, appropriately discounted to present value terms.

Therefore, the expected additional cost of retaining a tool instead of disposing of it is given by

$$R_0 + p \times \sum_{i=m}^{N} w_i S_i + (1-p) \times (S_N + D_N) - D_0. \tag{A.1}$$

The first term in equation A.1, R_0, is the nonrecurring cost of retaining the tool. The second term,

$$p \times \sum_{i=m}^{N} w_i S_i,$$

is the probability of a restart multiplied by expected recurring storage costs, conditional on a restart occurring. The third term, $(1 - p) \times (S_N + D_N)$, is the probability of no restart multiplied by the N years of storage costs and disposal costs in year N implied by no restart. Finally, we subtract D_0 because the decision to retain a tool implies there is no up-front disposal cost.

Next we turn to the expected cost avoidance associated with tool retention. In the event a retained tool is useful for a restart i years from now, the benefit it provides would be $A_i - U_i$, where A_i would be the cost of acquiring a new version of this tool in year i and U_i denotes the cost of preparing the tool for reuse, e.g., retrieving it from storage, transportation, refurbishment, and reinstallation.

If restart occurs with probability p, the expected cost avoidance of tool retention would be

$$p \times \sum_{i=m}^{N} w_i (A_i - U_i). \tag{A.2}$$

Tooling retention is desirable if expected cost avoidance (equation A.2) is greater than expected additional costs (equation A.1). A tool should be retained if and only if

$$p \times \sum_{i=m}^{N} w_i (A_i - U_i) - R_0 - p \times \sum_{i=m}^{N} w_i S_i - (1-p) \times (S_N + D_N) + D_0 > 0. \tag{A.3}$$

Figure A.1 is analogous to Figure 4.1, except that it uses this appendix's notation, e.g., up-front storage cost R_0, disposal costs D_0 and D_N, new tool cost A_i. We do not need to parameterize eventual disposal cost conditional on restart because that term cancels inequality A.3's subtraction.

Inequality A.3 can be solved for the breakeven value of p_{BE}, the probability of eventual restart that makes a decisionmaker indifferent between retaining tools and disposing of them immediately. This breakeven value of p_{BE} is

$$p_{BE} = \frac{R_0 + S_N + D_N - D_0}{\displaystyle\sum_{i=m}^{N} w_i(A_i - U_i - S_i) + S_N + D_N}. \tag{A.4}$$

p_{BE} is the minimum probability of future tool use for which tool retention is preferred to immediate tool disposal. If a decisionmaker's perceived probability of future tool use is greater than p_{BE}, the tool should be retained. If the decisionmaker's perceived probability of future tool use is less than p_{BE}, the tool should be disposed of. The lower the value of p_{BE}, the more desirable tool retention is.

We refer to the numerator of equation A.4 as the "differential cost of tool retention." This is the nonrecurring cost of tool retention plus the recurring cost of retaining tools for N years, then disposing of them, less the costs of initial disposal. The denominator of equation A.4 is the expected payoff associated with tool retention in the event of a restart.

Equation A.4 provides insight into the desirability of tooling retention. For example, increasing the nonrecurring costs of putting tools into storage (R_0) increases p_{BE}, making tooling retention less desirable. Higher costs for acquiring new tools (A_i) reduce p_{BE}, making tool retention more desirable. A shorter break between shutdown and restart decreases storage costs S_i, decreasing p_{BE}, making tool retention more desirable. Increasing stored tool reuse costs (U_i) increases p_{BE}, making tooling retention less desirable.

Figure A.1
A Tool Retention Decision Tree with Model Parameters

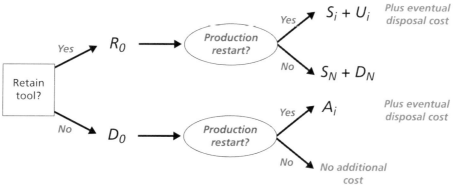

A Comparison of C-17A and F-22 Tooling Retention

In 2010, the Air Force decided to store F-22 production-only tooling at Sierra Army Depot using the Conex storage concept discussed in Chapter Two.

Retaining C-17A production-only tooling appears to be less desirable than retaining F-22 production-only tooling. Graser et al. (2011) indicated that F-22 production-only tooling could be stored at 9 cents for each dollar spent on acquiring that tooling. The methodologically comparable estimate for retaining all C-17A production-only tooling is 21 cents on the dollar. The C-17A is a much larger aircraft than the F-22, with larger, bulkier tools that are less valuable on a per-square-foot-stored basis, thus driving up the breakeven restart probability.

However, as discussed in Chapter Four, we do not recommend retaining all production-only C-17A tooling. Unlike the F-22 report, which treated its production-only tools as a single population, the category-by-category C-17A approach discussed in this report allows retention decisions to be tailored to different perceived restart probabilities.

There are additional factors, not accounted for in our breakeven calculations, that further discourage C-17A tooling retention relative to the F-22 case. At the time of the F-22 study, sustainment requirements had not been finalized, with the F-22 fleet having accumulated fewer than 100,000 flying hours, considered a major milestone in terms of determining sustainment requirements. As noted, the C-17A has over 2 million flying hours, so its sustainment requirements are much better known.

Also, contractor-financed equipment is more important in the C-17A case, so there would be more tooling gaps to fill to resume C-17 production, even if all government-funded production-only tools were retained.

Additionally, F-22 production could be restarted in Fort Worth, Texas, and Marietta, Georgia, allowing access to knowledgeable former F-22 production workers still employed in other programs in those facilities. By contrast, a C-17 restart using the Long Beach plant seems unlikely, with the likely loss of the entire workforce after production shutdown.

Finally, F-22 production technology will be just over ten years old at production shutdown; C-17A production technology is about 20 years old now. Hence, reconstituting the C-17A production line in its current form seems less likely than reconstituting that of the F-22.

Bibliography

Air Force Cost Analysis Agency, *C-5D CCA Sept. 95 Production Estimate*, September 1995.

Air Force Materiel Command Instruction 23-103, *Diminishing Manufacturing Sources and Materiel Shortages (DMSMS) Program*, October 13, 2000.

AirLaunch, LLC, "Operational C-17A Used to Break Another Record with AirLaunch in DARPA/Air Force Falcon Small Launch Vehicle Program," Kirkland, Wash.: AirLaunch LLC, July 27, 2006. As of April 11, 2011:
http://www.airlaunchllc.com/AirLaunch%20July%20Drop%20Test%20Press%20Release%20-%20Final%20072706%20with%20Pics.pdf

Birkler, John, and Joseph P. Large, *A Methodology for Estimating the Cost of Aircraft Structural Modification*, Santa Monica, Calif.: RAND Corporation, R-2565-AF, 1981. As of April 11, 2011:
http://www.rand.org/pubs/reports/R2565.html

Birkler, John, Joseph P. Large, Giles K. Smith, and Fred Timson, *Reconstituting a Production Capability: Past Experience, Restart Criteria, and Suggested Policies*, Santa Monica, Calif.: RAND Corporation, MR-273-ACQ, 1993. As of April 11, 2011:
http://www.rand.org/pubs/monograph_reports/MR273.html

Boeing Company, *C-17 Production Line Shutdown/Restart Technical Report*, Report No. NA-06-1511, November 3, 2006.

Capaccio, Tony, "Lockheed F-35's Projected Cost Now $382 Billion, Up 65 Percent," *Bloomberg Businessweek*, June 1, 2010.

Desjardins, Maj Gen Susan Y., Director, Strategic Plans, Requirements and Programs, Headquarters Air Mobility Command, "The Cost Effectiveness of Procuring Weapon Systems in Excess of Requirements," statement before the Senate Committee on Homeland Security and Government Affairs, Subcommittee on Federal Financial Management, Government Information, Federal Services, and International Security, July 13, 2010. As of April 11, 2011:
http://hsgac.senate.gov/public/index.cfm?FuseAction=Hearings.Hearing&Hearing_ID=82233a71-845c-4d44-a755-f58684ac6bb8

DoC—*See* U.S. Department of Commerce.

DoD—*See* U.S. Department of Defense.

Ebert, Lee G., *Special Tooling Disposition for Aircraft Entering Post Production Support*, master's thesis, Monterey, Calif.: Naval Postgraduate School, December 1992. As of April 11, 2011:
http://www.dtic.mil/cgi-bin/GetTRDoc?Location=U2&doc=GetTRDoc.pdf&AD=ADA261614

Galorath Incorporated, "SEER for Hardware, Electronics & Systems," 2011. As of March 4, 2011:
http://www.galorath.com/index.php/products/hardware/C4/

Gates, Dominic, "A 'Prescient' Warning to Boeing on 787 Trouble," *Seattle Times*, February 5, 2011. As of March 4, 2011:
http://seattletimes.nwsource.com/html/sundaybuzz/2014125414_sundaybuzz06.html

Gembutsu Consulting LLC, "Lean Manufacturing Glossary, Definitions and Terms," 2009. As of June 2, 2011:
http://www.gembutsu.com/articles/leanmanufacturingglossary.html#m

Graser, John C., Kevin Brancato, Guy Weichenberg, Soumen Saha, and Akilah Wallace, *Retaining F-22A Tooling: Options and Costs*, Santa Monica, Calif.: RAND Corporation, TR-831-AF, 2011. As of April 11, 2011:
http://www.rand.org/pubs/technical_reports/TR831.html

Hartmann, John, Chris Meeker, Mike Weller, Nigel Izzard, Andrew Smith, Alan Ferguson, and Alan Ellson, "Determinate Assembly of Tooling Allows Concurrent Design of Airbus Wings and Major Assembly Fixtures," paper presented at the SAE 2004 Aerospace Manufacturing & Automated Fastening Conference & Exhibition SAE International, September 21, 2004. As of August 15, 2011:
http://papers.sae.org/2004-01-2832/

Lockheed Corporation, untitled file of proprietary cost data from the C-5B program, undated.

McCord, Mike, Principal Deputy Under Secretary of Defense (Comptroller), Statement before the Senate Committee on Homeland Security and Governmental Affairs, Subcommittee on Federal Financial Management, Government Information, Federal Services, and International Security, July 13, 2010.

"Military Containers," GlobalSecurity.org, undated. As of March 4, 2011:
http://www.globalsecurity.org/military/systems/ground/container.htm

Office of Management and Budget, "Regulatory Analysis," Circular A-4, to the heads of executive agencies and establishments, September 17, 2003. As of June 20, 2011:
http://www.whitehouse.gov/omb/circulars_a004_a-4

———, "Discount Rates for Cost-Effectiveness, Lease Purchase, and Related Analyses," Circular A-94 Appendix C, December 2010. As of February 16, 2011:
http://www.whitehouse.gov/omb/circulars_a094/a94_appx-c/

Posner, Richard A., "Catastrophic Risks, Resource Allocation, and Homeland Security," *Journal of Homeland Security*, October 2005. As of June 20, 2011:
http://www.homelandsecurity.org/journal/Default.aspx?oid=133&ocat=1&AspxAutoDetectCookieSupport=1

USAF —*See* U.S. Air Force.

U.S. Air Force, "C-17 Globemaster III," fact sheet, October 22, 2008. As of February 19, 2011:
http://www.af.mil/information/factsheets/factsheet.asp?id=86

———, *Committee Staff Procurement Backup Book Fiscal Year (FY) 2011 Budget Estimates, Aircraft Procurement Air Force*, Vol. II, Washington, D.C.: Office of the Secretary of the Air Force, February 2010.

U.S. Department of Commerce, Bureau of Industry and Security, Office of Strategic Industries and Economic Security, *National Security Assessment of the C-17 Globemaster Cargo Aircraft's Economic & Industrial Base Impacts: Final Report*, December 2005.

U.S. Department of Defense, *Selected Acquisition Report, C-5 AMP*, as of December 31, 2009a.

———, *Selected Acquisition Report, C-5 RERP*, as of December 31, 2009b.

———, *Selected Acquisition Report, Draft SAR, C-130 AMP*, as of June 30, 2010.

U.S. Department of Defense, Deputy Under Secretary of Defense for Installations and Environment, *DoD Facilities Pricing Guide for FY 2010*, UFC 3-701-01, June 2010.

U.S. Government Accountability Office, *Defense Acquisitions: Timely and Accurate Estimates of Costs and Requirements Are Needed to Define Optimal Future Strategic Airlift Mix*, GAO-09-50, November 2008.

York, James, Richard Krens, Krista Pezold, Brooke Sampey, Patti Jo Vore, and John L. Sullivan, *Cost of Production Line Shutdown and Related Activities*, Falls Church, Va.: MCR Services Group, TR-9615/01, 1996.

Younossi, Obaid, Kevin Brancato, John C. Graser, Thomas Light, Rena Rudavsky, and Jerry M. Sollinger, *Ending F-22A Production: Costs and Industrial Base Implications of Alternative Options*, Santa Monica, Calif.: RAND Corporation, MG-797-AF, 2010. As of April 11, 2011:
http://www.rand.org/pubs/monographs/MG797.html

Younossi, Obaid, Kevin Brancato, Fred Timson, and Jerry Sollinger, *Starting Over: Technical, Schedule, and Cost Issues Involved with Restarting C-2 Production*, Santa Monica, Calif.: RAND Corporation, MG-203-Navy, 2003, Not available to the general public.

Younossi, Obaid, Michael Kennedy, and John C. Graser, *Military Airframe Costs: The Effects of Advanced Materials and Manufacturing Processes*, Santa Monica, Calif.: RAND Corporation, MR-1370-AF, 2001. As of April 11, 2011:
http://www.rand.org/pubs/monograph_reports/MR1370.html